DATE

VOLUME FIVE HUNDRED AND SEVENTY ONE

METHODS IN ENZYMOLOGY

Rational Design of Enzyme-Nanomaterials

METHODS IN ENZYMOLOGY

Editors-in-Chief

ANNA MARIE PYLE
*Departments of Molecular, Cellular and Developmental
Biology and Department of Chemistry Investigator
Howard Hughes Medical Institute
Yale University*

DAVID W. CHRISTIANSON
*Roy and Diana Vagelos Laboratories
Department of Chemistry
University of Pennsylvania
Philadelphia, PA*

Founding Editors

SIDNEY P. COLOWICK and NATHAN O. KAPLAN

VOLUME FIVE HUNDRED AND SEVENTY ONE

METHODS IN
ENZYMOLOGY

Rational Design of
Enzyme-Nanomaterials

Edited by

CHALLA VIJAYA KUMAR

*Department of Chemistry,
Department of Molecular and Cell Biology,
University of Connecticut,
Storrs, Connecticut, USA*

AMSTERDAM • BOSTON • HEIDELBERG • LONDON
NEW YORK • OXFORD • PARIS • SAN DIEGO
SAN FRANCISCO • SINGAPORE • SYDNEY • TOKYO

Academic Press is an imprint of Elsevier

Academic Press is an imprint of Elsevier
50 Hampshire Street, 5th Floor, Cambridge, MA 02139, USA
525 B Street, Suite 1800, San Diego, CA 92101-4495, USA
The Boulevard, Langford Lane, Kidlington, Oxford OX5 1GB, UK
125 London Wall, London, EC2Y 5AS, UK

First edition 2016

Copyright © 2016 Elsevier Inc. All rights reserved.

No part of this publication may be reproduced or transmitted in any form or by any means, electronic or mechanical, including photocopying, recording, or any information storage and retrieval system, without permission in writing from the publisher. Details on how to seek permission, further information about the Publisher's permissions policies and our arrangements with organizations such as the Copyright Clearance Center and the Copyright Licensing Agency, can be found at our website: www.elsevier.com/permissions.

This book and the individual contributions contained in it are protected under copyright by the Publisher (other than as may be noted herein).

Notices
Knowledge and best practice in this field are constantly changing. As new research and experience broaden our understanding, changes in research methods, professional practices, or medical treatment may become necessary.

Practitioners and researchers must always rely on their own experience and knowledge in evaluating and using any information, methods, compounds, or experiments described herein. In using such information or methods they should be mindful of their own safety and the safety of others, including parties for whom they have a professional responsibility.

To the fullest extent of the law, neither the Publisher nor the authors, contributors, or editors, assume any liability for any injury and/or damage to persons or property as a matter of products liability, negligence or otherwise, or from any use or operation of any methods, products, instructions, or ideas contained in the material herein.

ISBN: 978-0-12-804680-7
ISSN: 0076-6879

For information on all Academic Press publications
visit our website at https://www.elsevier.com/

Publisher: Zoe Kruze
Acquisition Editor: Zoe Kruze
Editorial Project Manager: Sarah Lay
Production Project Manager: Magesh Kumar Mahalingam
Designer: Matthew Limbert

Typeset by SPi Global, India

CONTENTS

Contributors ix
Preface xiii

1. Preparation of Biocatalytic Microparticles by Interfacial Self-Assembly of Enzyme–Nanoparticle Conjugates Around a Cross-Linkable Core   1

S.M. Andler, L.-S. Wang, J.M. Goddard, and V.M. Rotello

1. Theory   2
2. Equipment   6
3. Materials   6
4. Protocol   7
5. Step 1: Nanoparticle Synthesis   7
6. Step 2: Purification of enzyme   10
7. Step 3: Preparation of the Aqueous Phase and Oil Phase   12
8. Step 4: Microparticle Assembly   14
9. Step 5: Microparticle Washing   15
10. Conclusions   16
References   16

2. Monitoring Enzymatic Proteolysis Using Either Enzyme- or Substrate-Bioconjugated Quantum Dots   19

S.A. Díaz, J.C. Breger, and I.L. Medintz

1. Introduction   20
2. Quantification Assay for Observing Modified Kinetics with Enzyme–QD Conjugates   24
3. Enzyme Activity Sensors Based on Transient QD–Enzyme Interactions   34
4. Notes   49
Acknowledgments   50
References   50

3. Intense PEGylation of Enzyme Surfaces: Relevant Stabilizing Effects   55

S. Moreno-Pérez, A.H. Orrego, M. Romero-Fernández, L. Trobo-Maseda, S. Martins-DeOliveira, R. Munilla, G. Fernández-Lorente, and J.M. Guisan

1. Introduction   56
2. Theory   57

3. Protocols	62
4. Inactivation of Modified Enzyme Derivatives	68
5. Conclusions	70
Acknowledgments	71
References	71

4. Immobilization of Lipases on Heterofunctional Octyl–Glyoxyl Agarose Supports: Improved Stability and Prevention of the Enzyme Desorption 73

N. Rueda, J.C.S. dos Santos, R. Torres, C. Ortiz, O. Barbosa, and R. Fernandez-Lafuente

1. Theory	74
2. Equipment	76
3. Materials	76
4. Step 1. Preparation of the Support Octyl–Glyoxyl Agarose	77
5. Step 2. Immobilization of Lipases via Interfacial Activation on Octyl–Glyoxyl Agarose	79
6. Step 3. Covalent Immobilization of Adsorbed Lipases on Octyl–Glyoxyl Agarose	82
Acknowledgments	84
References	84

5. Biomimetic/Bioinspired Design of Enzyme@capsule Nano/Microsystems 87

J. Shi, Y. Jiang, S. Zhang, D. Yang, and Z. Jiang

1. Introduction	88
2. General Procedure of the Design and Construction of Enzyme@capsule Nano/Microsystems Through Biomimetic/Bioinspired Methods	93
3. Some Specific Examples	97
4. Concluding Remarks	108
Acknowledgment	110
References	110

6. Synergistic Functions of Enzymes Bound to Semiconducting Layers 113

K. Kamada, A. Yamada, M. Kamiuchi, M. Tokunaga, D. Ito, and N. Soh

1. Introduction	114
2. Fabrication of Enzyme-Intercalated Layered Oxides	116

3. Activity of Enzymes Bound to Titanate Layers　123
　　4. Photochemical Control of Enzymatic Activity of Oxidoreductases
　　　　Bound to Layered Oxides　125
　　5. Biorecognition Using Doped Titanate Layers Modified
　　　　with Biomolecules　129
　　6. Magnetic Application of Hybrids Composed of Enzymes
　　　　and Doped Titanates　130
　　7. Conclusions　132
　　Acknowledgments　133
　　References　133

7. Bioconjugation of Antibodies and Enzyme Labels onto Magnetic Beads　135
B.A. Otieno, C.E. Krause, and J.F. Rusling

　　1. Introduction　136
　　2. Bioconjugation of Magnetic Beads　138
　　3. Characterization of Magnetic Bead Bioconjugates　142
　　4. Integration of Magnetic Beads into Immunoassay　146
　　Acknowledgment　148
　　References　148

8. Rationally Designed, "Stable-on-the-Table" NanoBiocatalysts Bound to Zr(IV) Phosphate Nanosheets　151
I.K. Deshapriya and C.V. Kumar

　　1. Introduction　152
　　2. Methods　160
　　References　174

9. Portable Enzyme-Paper Biosensors Based on Redox-Active CeO_2 Nanoparticles　177
A. Karimi, A. Othman, and S. Andreescu

　　1. Introduction　178
　　2. NPs-Based Enzyme Biosensors　179
　　3. CeO_2 NPs for Enzyme Immobilization and Enzyme-Based Biosensors　183
　　4. Design of a CeO_2-Based Colorimetric Enzyme Biosensor　185
　　5. Comments on the Method　190
　　Acknowledgments　192
　　References　192

10. Rational Design of Nanoparticle Platforms for "Cutting-the-Fat": Covalent Immobilization of Lipase, Glycerol Kinase, and Glycerol-3-Phosphate Oxidase on Metal Nanoparticles — 197

V. Aggarwal and C.S. Pundir

1. Introduction — 198
2. Use of Rationally Designed Nanoscaffolds for Enzyme Binding — 200
3. Use of Chitosan for Enhancing Nanoparticle Surface Chemistry — 201
4. Experimental — 202
References — 222

11. BioGraphene: Direct Exfoliation of Graphite in a Kitchen Blender for Enzymology Applications — 225

C.V. Kumar and A. Pattammattel

1. Introduction — 226
2. Mechanism of Exfoliation — 230
3. Tunability of the BioGraphene Characteristics — 231
4. Protein Binding to Graphene and Some Biological Applications — 232
5. Methods — 233
6. Conclusions — 241
Acknowledgments — 242
References — 242

Author Index — *245*
Subject Index — *257*

CONTRIBUTORS

V. Aggarwal
Department of Biochemistry, Maharshi Dayanand University, Rohtak, Haryana, India

S.M. Andler
Department of Food Science, University of Massachusetts-Amherst, Amherst, Massachusetts, USA

S. Andreescu
Department of Chemistry and Biomolecular Science, Clarkson University, Potsdam, New York, USA

O. Barbosa
Departamento de Química, Facultad de Ciencias, Universidad del Tolima, Ibagué, Colombia

J.C. Breger
Center for Bio/Molecular Science and Engineering, Code 6900, U.S. Naval Research Laboratory; American Society for Engineering Education, Washington, District of Columbia, USA

S.A. Díaz
Center for Bio/Molecular Science and Engineering, Code 6900, U.S. Naval Research Laboratory; American Society for Engineering Education, Washington, District of Columbia, USA

I.K. Deshapriya
Department of Chemistry, University of Connecticut, Storrs, Connecticut, USA

J.C.S. dos Santos
Departamento de Biocatálisis, Instituto de Catálisis-CSIC, Campus UAM-CSIC, Madrid, Spain; Departamento de Engenharia Química, Universidade Federal Do Ceará, Campus Do Pici, Fortaleza, Ceará, Brazil

R. Fernandez-Lafuente
Departamento de Biocatálisis, Instituto de Catálisis-CSIC, Campus UAM-CSIC, Madrid, Spain

G. Fernández-Lorente
Institute of Catalysis, Spanish Research Council, CSIC, Madrid, Spain

J.M. Goddard
Department of Food Science, University of Massachusetts-Amherst, Amherst, Massachusetts, USA

J.M. Guisan
Institute of Catalysis, Spanish Research Council, CSIC, Madrid, Spain

D. Ito
Graduate School of Engineering, Nagasaki University, Nagasaki, Japan

Y. Jiang
School of Chemical Engineering and Technology, HeBei University of Technology, Tianjin, China

Z. Jiang
Collaborative Innovation Center of Chemical Science and Engineering (Tianjin); Key Laboratory for Green Chemical Technology of Ministry of Education, School of Chemical Engineering and Technology, Tianjin University, Tianjin, China

K. Kamada
Graduate School of Engineering, Nagasaki University, Nagasaki, Japan

M. Kamiuchi
Graduate School of Engineering, Nagasaki University, Nagasaki, Japan

A. Karimi
Department of Chemistry and Biomolecular Science, Clarkson University, Potsdam, New York, USA

C.E. Krause
Department of Chemistry, University of Hartford, West Hartford, Connecticut, USA

C.V. Kumar
Department of Chemistry; Department of Molecular and Cell Biology; Institute of Material Science, University of Connecticut, Storrs, Connecticut, USA; Department of Inorganic and Physical Chemistry, Indian Institute of Science, Bengaluru, Karnataka, India

S. Martins-DeOliveira
Institute of Catalysis, Spanish Research Council, CSIC, Madrid, Spain

I.L. Medintz
Center for Bio/Molecular Science and Engineering, Code 6900, U.S. Naval Research Laboratory, Washington, District of Columbia, USA

S. Moreno-Pérez
Institute of Catalysis, Spanish Research Council, CSIC, Madrid, Spain

R. Munilla
Institute of Catalysis, Spanish Research Council, CSIC, Madrid, Spain

A.H. Orrego
Institute of Catalysis, Spanish Research Council, CSIC, Madrid, Spain

C. Ortiz
Escuela de Bacteriología y Laboratorio Clínico, Universidad Industrial de Santander, Bucaramanga, Colombia

A. Othman
Department of Chemistry and Biomolecular Science, Clarkson University, Potsdam, New York, USA

B.A. Otieno
Department of Chemistry, University of Connecticut, Storrs, Connecticut, USA

A. Pattammattel
Department of Chemistry, University of Connecticut, Storrs, Connecticut, USA

C.S. Pundir
Department of Biochemistry, Maharshi Dayanand University, Rohtak, Haryana, India

M. Romero-Fernández
Institute of Catalysis, Spanish Research Council, CSIC, Madrid, Spain

V.M. Rotello
Department of Chemistry, University of Massachusetts-Amherst, Amherst, Massachusetts, USA

N. Rueda
Departamento de Biocatálisis, Instituto de Catálisis-CSIC, Campus UAM-CSIC, Madrid, Spain; Escuela de Química, Grupo de investigación en Bioquímica y Microbiología (GIBIM), Edificio Camilo Torres 210, Universidad Industrial de Santander, Bucaramanga, Colombia

J.F. Rusling
Department of Chemistry; Institute of Materials Science, University of Connecticut, Storrs; Department of Cell Biology, University of Connecticut Health Center, Farmington, Connecticut, USA; School of Chemistry, National University of Ireland, Galway, Ireland

J. Shi
School of Environmental Science and Engineering, Tianjin University; Collaborative Innovation Center of Chemical Science and Engineering (Tianjin), Tianjin, China

N. Soh
Faculty of Agriculture, Saga University, Saga, Japan

M. Tokunaga
Graduate School of Engineering, Nagasaki University, Nagasaki, Japan

R. Torres
Escuela de Química, Grupo de investigación en Bioquímica y Microbiología (GIBIM), Edificio Camilo Torres 210, Universidad Industrial de Santander, Bucaramanga, Colombia

L. Trobo-Maseda
Institute of Catalysis, Spanish Research Council, CSIC, Madrid, Spain

L.-S. Wang
Department of Chemistry, University of Massachusetts-Amherst, Amherst, Massachusetts, USA

A. Yamada
Graduate School of Engineering, Nagasaki University, Nagasaki, Japan

D. Yang
School of Environmental Science and Engineering, Tianjin University; Collaborative Innovation Center of Chemical Science and Engineering (Tianjin), Tianjin, China

S. Zhang
Key Laboratory for Green Chemical Technology of Ministry of Education, School of Chemical Engineering and Technology, Tianjin University, Tianjin, China

PREFACE

Enzymes are nature's catalysts and they are responsible for accelerating most chemical reactions in biology. The high selectivity or specificity of these catalysts, which operate under mild conditions of biological systems, makes them very attractive for use in the laboratory or industry. But this is quite challenging, because enzymes have evolved to function in the complex biological media at biological temperatures, and hence their stability/function under nonbiological conditions of a common chemical laboratory such as organic solvent media, extreme pHs, or high temperatures is often poor. Most enzymes are also quite expensive when compared to many chemical reagents, and hence special methods are required to render enzymes usable in the context of a chemical laboratory or for industry. This volume addresses these two important issues related to enzymes in much detail, and builds on the important contributions made to this topic from previous volumes of this series.[1]

Recent developments in nanochemistry are brought to bear on the above two problems of enzyme stability/function in the current volume. From a fundamental chemical point of view, enzyme stability is a thermodynamic issue, while kinetic stability is also important. Enzymes are thermodynamically stable at room temperature but the denaturation free energy (ΔG) decreases with increase in temperature. It reaches a numerical value of zero at the denaturation temperature, and at this temperature, the native and denatured states of the enzyme are in equilibrium with equal concentrations. Above this temperature, the denatured state is thermodynamically preferred over the native state. Similarly, the pH and solvent dielectric can also influence ΔG and render enzymes unstable. To combat this important issue, significant progress has been made in improving enzyme stability by reducing the conformational entropy of enzymes by encasing them in nanomaterials with a variety of architectures described in this volume. One important point is that entropy of denaturation (ΔS) is positive and any reduction in ΔS raises ΔG and hence contributes to the enhanced stability of the enzyme. Many methods are developed to take advantage of this thermodynamic principle, directly or indirectly to lower the conformational entropy of the denatured state, and nanochemistry methods have become

---

[1] Methods in Enzymology Volumes 137, 136, 135, 64, and 44.

handy for this approach. Our understanding of these issues, therefore, is rooted in fundamentals so that the tree of knowledge can grow and prosper. This is what the authors of the chapters in this volume have striven to do.

Nanomaterials have several unique properties, such as high surface area for unit mass, rich surface chemistry arising from surface atoms which differ significantly from those of the interior atoms, and these materials also have versatile optical, thermal, and magnetic properties. When these interesting properties of nanomaterials are coupled with the versatile catalytic abilities of enzymes, one can generate novel biocatalytic nanomaterials. Many examples are documented in this volume. Very high loadings of the enzymes on the nanoparticle are often achieved due to their large surface-to-mass ratio. Due to the nanosize of the support materials, diffusion of the substrate to the active site of the bound enzyme is also facilitated.

One major difficulty in using nanomaterials for enzyme binding and recycling, however, is that the separation of the nanoparticle bound enzyme from the reaction media is often difficult due to the dispersion of the particles in the solvent. This important issue has been addressed by using several different methods in this volume, and one of them is the use of magnetic nanoparticles, and the other has been using innovative design architectures where the nanoparticles themselves are embedded in porous microparticles that could be readily separated from the reaction media and recycled (Fig. 1). Easy separation, reaction workup, and recycling of the biocatalyst could reduce the process cost and hence very important for industrial applications. This kind of hierarchical structure design from the nanoscale to macroscale with attention to the molecular details of the support matrix and its porous nature is critical for rapid progress in this area. This was the subject matter of several distinguished chapters in this volume.

The use of two-dimensional materials (nanosheets) for enzyme binding is another major breakthrough in the field. Because of their nm-thinness and large lateral size of several microns, nanosheets have very large surface-area-to-mass ratios and large ratio of length to thickness (aspect ratio) greater than their nanoparticle cousins. Therefore, large amounts of enzymes can be bound per gram of these nanosheets, and due to this unique aspect ratio, the bound enzymes are well exposed to the solvent for rapid reactions, overcoming the diffusional issues related to enzymes bound on nanoparticle surfaces. Furthermore, these nanosheets can be restacked with enzymes entrapped between the plates, and the nanogalleries formed by restacking provide protection against proteases, bacteria, and viruses, which can potentially degrade the biocatalyst. When properly designed,

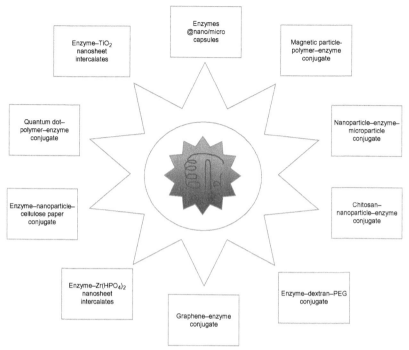

Figure 1 The variety of nanoarchitectures and methods used in the design of enzyme–nanomaterial hybrids for advanced enzymology applications.

enzyme/nanosheet biocatalysts also enhance thermal stabilities of bound enzymes due to a significant reduction in the conformational entropy of the bound enzyme. This latter hypothesis of confinement and reduced entropy of enzymes embedded between nanosheets (two-dimensional space) is now being tested by a number of different research groups around the world.

A number of two-dimensional materials are also being explored for enzyme binding, including the state-of-the-art materials such as graphene, which could be beneficial for enzyme binding due to its unique properties. For example, charge-conducting nanosheets might be able to promote redox activities of enzymes by wiring the charge-donor and charge-acceptor sites, via the layered material. This hypothesis remains to be tested, and facile access to two-dimensional materials reported in this volume might accelerate such exciting studies.

Cellulosic paper continues to attract attention for enzyme binding and sensing applications due to its versatile features as well as low cost. Paper might be the medium of the future for enzyme binding where the nanofibers of cellulose could be exploited for functionalization as well as physical interlocking of enzymes in the nanofibers. Paper-based sensors, as reported here and elsewhere, are very exciting for the design of light weight, inexpensive, porous, high-capacity bioreactors or enzyme cartridges for small- or large-scale synthesis applications. One other area that could benefit from these types of developments could be synthetic organic chemistry where a multistep synthesis can be carried out by a number of different enzymes embedded in the layers of a paper matrix. Such futuristic goals are not far away, and numerous examples of paper-embedded enzymes are already known and it is only a matter of systematic studies that would one day produce reaction cartridges for multistep synthesis of high value or tedious chemical reactions.

Industrial exploitation of nature's biocatalysts is still rudimentary, when one compares the current status of the field with the potential number of enzymes that have not yet been examined for attachment to nanosurfaces. From nanoparticles to nanomicrocomposites to two-dimensional materials to fibrous networks (paper), there appears to be limitless opportunities to exploit and control the behavior of enzymes under nonbiological conditions. Enzymes can influence the production of molecules, materials, and devices, and they are at the service of the mankind on a global scale. Therefore, they have the potential to make progress in addressing social, economical, environmental, and political issues of the current-day society.

ACKNOWLEDGMENTS

Author thanks the Fulbright Foundation for a fellowship to visit the Indian Institute of Science, India, and the National Science Foundation for partial financial support of this work (DMR-1401879).

CHALLA VIJAYA KUMAR

Department of Chemistry and Department of Molecular and Cell Biology,
University of Connecticut, Storrs, Connecticut, USA
Inorganic and Physical Chemistry Department, Indian Institute of Science,
Bengaluru, Karnataka, India

CHAPTER ONE

Preparation of Biocatalytic Microparticles by Interfacial Self-Assembly of Enzyme–Nanoparticle Conjugates Around a Cross-Linkable Core

S.M. Andler*, L.-S. Wang[†], J.M. Goddard*, V.M. Rotello[†,1]

*Department of Food Science, University of Massachusetts-Amherst, Amherst, Massachusetts, USA
[†]Department of Chemistry, University of Massachusetts-Amherst, Amherst, Massachusetts, USA
[1]Corresponding author: e-mail address: rotello@chem.umass.edu

Contents

1. Theory — 2
2. Equipment — 6
3. Materials — 6
   3.1 Buffer Preparation — 7
4. Protocol — 7
   4.1 Duration — 7
   4.2 Preparation — 7
5. Step 1: Nanoparticle Synthesis — 7
   5.1 Overview — 7
   5.2 Duration — 8
   5.3 Tip — 10
   5.4 Tip — 10
   5.5 Tip — 10
   5.6 Tip — 10
   5.7 Tip — 10
6. Step 2: Purification of enzyme — 10
   6.1 Overview — 10
   6.2 Duration — 11
   6.3 Tip — 11
   6.4 Tip — 12
7. Step 3: Preparation of the Aqueous Phase and Oil Phase — 12
   7.1 Overview — 12
   7.2 Duration — 12
   7.3 Tip — 12
   7.4 Tip — 13
   7.5 Tip — 13

Methods in Enzymology, Volume 571
ISSN 0076-6879
http://dx.doi.org/10.1016/bs.mie.2016.02.005

© 2016 Elsevier Inc.
All rights reserved.

7.6 Tip 13
7.7 Tip 13
7.8 Tip 13
7.9 Tip 13
8. Step 4: Microparticle Assembly 14
   8.1 Overview 14
   8.2 Duration 14
   8.3 Tip 14
   8.4 Tip 15
   8.5 Tip 15
9. Step 5: Microparticle Washing 15
   9.1 Overview 15
   9.2 Duration 15
   9.3 Tip 16
   9.4 Tip 16
10. Conclusions 16
References 16

Abstract

Rational design of hierarchical interfacial assembly of reusable biocatalytic microparticles is described in this chapter. Specifically, purified enzymes and functionalized nanoparticles are electrostatically assembled at the interface of cross-linked microparticles which are formed through ring opening metathesis polymerization. The diameters of microparticle assemblies average 10 μm, and they show enhanced kinetic efficiency as well as improved stability against heat, pH, and solvent denaturation when compared to stabilities of the corresponding native enzymes.

1. THEORY

Enzyme-catalyzed reactions offer enhanced specificity and selectivity over traditional catalysts, but in the native form, many enzymes suffer poor solvent stability and thermostability. To improve their performance and ease of recovery in industrial applications, enzymes have been immobilized on solid supports (Talbert & Goddard, 2012). Immobilization has been achieved through numerous methods including covalent attachment, physical adsorption, cross-linking, self-assembly, and entrapment (Hwang & Gu, 2013; Sheldon & Pelt, 2013). Physical adsorption achieves high-protein loading on a surface with minimal activity losses. The main drawback to physical adsorption is the leaching of enzyme from the solid support, thus leading to decreased activity over multiple uses (Bastida et al., 1998; Jun et al., 2013; Wiemann et al., 2009). Enzymes can be covalently bound to solid supports or to other enzymes to form cross-linked aggregates (Sheldon, 2011) and covalent attachment prevents leaching of the enzyme during the catalytic cycles, but cross-linkers can alter enzyme conformation, and the chemistry used for linking may damage the delicate active site residues and thus reduce enzymatic activity of the resulting biocatalyst (Arroyo, Sánchez-Montero, & Sinisterra, 1999; Poppe, Costa, Brasil, Rodrigues, & Ayub, 2013). Entrapment or encapsulation of enzymes in inert matrices can facilitate the incorporation of enzymes into larger materials. The core material used to entrap the enzyme may inhibit substrate interaction with the active site of the enzyme, or enzymes may escape the matrix if the pores of the encapsulating matrix are too big (Yagonia, Park, & Yoo, 2013). Nanomaterials offer increased surface area to volume ratio, enabling increased protein loading and reduced protein–protein and protein–surface interactions that may result in activity loss (Jeong, Duncan, Park, Kim, & Rotello, 2011; Talbert et al., 2014). However, practical challenges such as recovery of the nanoparticle and their toxicity concerns have limited commercial adoption of biocatalytic nanomaterials in some industries.

An ideal immobilized enzyme system has been described as one in which the enzyme retains active conformation, catalytic activity, and substrate accessibility and does not leach from the solid support (Adlercreutz, 2013). We describe a method of enzyme immobilization and stabilization in which enzyme–nanoparticle complexes self-assemble at the interface of a cross-linkable oil core (Fig. 1) surrounded by water. Lipase B from *Candida antarctica* is used as a model enzyme to demonstrate this protocol. Lipase is a hydrolase that can catalyze esterification reactions with high specificity (Adlercreutz, 2013), and it has industrial application in biofuel production and food processing (Hasan, Shah, & Hameed, 2006). However, native lipases are sensitive to the conditions utilized in processing, such as elevated temperatures and organic solvents. Stabilized lipase microparticles are prepared as follows. Lipase is mixed with dopamine-capped iron oxide core nanoparticles to form electrostatically assembled enzyme–nanoparticle complexes in which nanoparticles stabilize enzyme conformation and enable magnetic recovery of the biocatalyst from the reaction mixture by simply applying a magnetic field. Enzyme–nanoparticle complexes are then mixed with a 10% (v/v) dicyclopentadiene/trichlorobenzene oil phase containing Grubbs' catalyst and 90% (v/v) aqueous phase (Bang, Mohite, Malik, & Nicholas, 2015). The oil–water mixture is emulsified, and during this, the enzyme–nanoparticle complexes self-assemble at the interface of water and microparticles. Since the enzymes reside at this interface, accessibility of the enzyme active sites to the substrates is enhanced and ensured

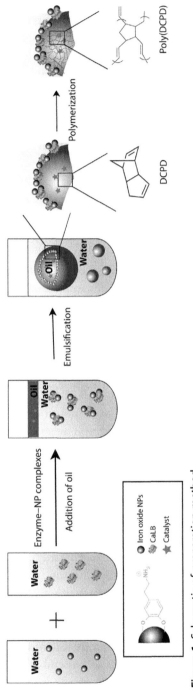

Figure 1 Schematic of preparation method.

(Fig. 1). On the other hand, polymerization of the oil core in the presence of catalyst results in a mechanically robust immobilized biocatalytic system (Fig. 2), which actively prevents leaching of the enzyme from the particles during the catalytic cycle or storage of the biocatalyst for future use. Michaelis–Menten kinetics of these systems demonstrated improved catalytic efficiency of stabilized lipase when compared to those of the native lipase. Additionally, substrate affinity for the enzyme active site is fully maintained after the assembly, which demonstrates the absence of any mass transfer limitations (Talbert et al., 2014), which are otherwise common with immobilized enzymes.

Preparation of the biocatalytic polymer microparticles can be adapted to alternate layers of enzymes and core materials (Jeong et al., 2011; Samanta et al., 2009). Nanoparticle:enzyme molar ratios can be altered to tailor the assemblies of enzyme–nanoparticle complexes according to enzyme isoelectric point. Ligands capping the nanoparticle surfaces can be selected to manipulate the interfacial microenvironment between the particle and the solvent to improve biocatalytic performance at the target pH values suitable for optimum biocatalyst performance. The coatings on the particles may also modulate the solubilization or sequestration of the substrate and further allow for rational manipulation of the biocatalytic kinetics. Magnetic properties can be induced in the nanoparticles in order to improve the recovery efficiency of the biocatalyst for easy workup of the reaction mixture and for recycling the biocatalyst (Dyal et al., 2003). These features are important for economic aspects of the biocatalyst and eventual environmental fate of the biocatalytic materials.

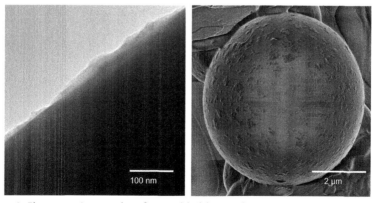

Figure 2 Electron micrographs of assembled biocatalytic microparticle. Left, TEM of edge of microparticle at the microparticle interface; right, SEM of microparticle.

Furthermore, the core material in the above design can be selected to enable better performance even in nonaqueous solvent systems which might be more suitable for certain hydrophobic substrates or when water is not desirable in the reaction medium. This rational design of hierarchical interfacial assembly of the biocatalyst at multiple levels is a major advance in controlling the biocatalyst properties at the molecular level. It is also a facile method to create reusable biocatalysts with dramatically improved stability against temperature, pH, and solvent-induced denaturation, as demonstrated here.

2. EQUIPMENT

Three-neck round-bottom flask, electronic balance, sonicator bath, refluxing apparatus, nitrogen gas, heating mantle, temperature controller, thermocouple, magnetic stir bar
Stirrer, centrifuge, 50-ml centrifuge tubes, volumetric flask, microcentrifuge, micropipettors
Pipettor tips, 1.5-ml polypropylene microcentrifuge tubes, amalgamator (e.g., Wykle Research, 4200 cpm), 0.5-ml 10K MWCO centrifugal filter tubes, 0.22-μm syringe filter
5-ml Luer-Lock syringe, mini-tube rotator, plate reader, UV-compatible 96-well plates, Zetasizer Nano ZS (Malvern Instruments)

3. MATERIALS

Lipase CaLB (*C. antarctica* lipase B) (e.g., ChiralVision, CV-CALBY)
Diethylene glycol (DEG) (e.g., Alfa Aesar, Cat. # A14728)
Sodium hydroxide (NaOH)
Iron(III) chloride hexahydrate ($FeCl_3 \cdot 6H_2O$) (e.g., Acros Organics, Cat. # 125030010)
Hydroxytyramine hydrochloride (Dopamine) (e.g., Acros Organics, Cat. # 122000100)
Iron(II) chloride tetrahydrate ($FeCl_2 \cdot 4H_2O$) (e.g., Sigma-Aldrich, Cat. # 44939)
Dicyclopentadiene (DCPD) (e.g., Acros Organics, Cat. #150761000)
Ruthenium-based first-generation Grubbs' catalyst (e.g., Sigma-Aldrich, Cat. # 579726)
1,2,4-Tricholorbenzene (TCB) (e.g., Acros Organics, Cat. # 157900010)
Ethanol, sodium acetate, acetic acid, toluene, MES hydrate, MES sodium salt

Table 1 MES Buffer

Component	Final Concentration (mM)	Stock (mM)	Volume (ml)
MES sodium salt	10	100	28
MES hydrate	10	100	22

Table 2 Sodium Acetate Buffer

Component	Final Concentration (mM)	Stock (mM)	Volume (ml)
Sodium acetate	10	100	10
Distilled water			85

3.1 Buffer Preparation

See Table 1.

Mix both buffer components and adjust pH to 7.0 using NaOH solution. See Table 2.

Using a 100-ml volumetric flask, mix both components and adjust pH to 5.0 using acetic acid. Fill up to 100 ml mark with distilled water.

4. PROTOCOL

4.1 Duration

Protocol: About 12.5 h
 See Fig. 3.

4.2 Preparation

This protocol may be scaled up or down, depending on the desired sample size. Additional microcentrifuge tubes can be added for greater number of samples, and the number of samples must be taken into consideration when preparing purified lipase and the oil phase required.

5. STEP 1: NANOPARTICLE SYNTHESIS

5.1 Overview

Dopamine-capped iron oxide nanoparticles were made by one-pot procedure using DEG as the solvent (Qu, Caruntu, Liu, & O'Connor, 2011). A schematic of nanoparticle synthesis is given in Fig. 4.

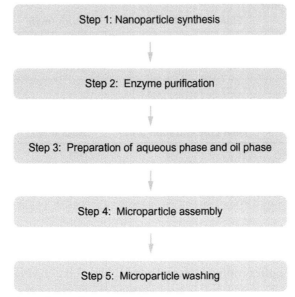

Figure 3 Flowchart of complete protocol.

5.2 Duration

6 h

- **1.1** Weigh 99.4 mg of $FeCl_2 \cdot 4H_2O$ (0.5 mmol) and 270.3 mg of $FeCl_3 \cdot 6H_2O$ (1 mmol) and mix with 20 g DEG.
- **1.2** Sonicate until iron salts dissolved in DEG and then transfer into a three-neck round-bottom flask for refluxing.
- **1.3** Set up the refluxing apparatus and then purge with nitrogen gas for 5 min to remove air present in the apparatus.
- **1.4** Dissolve 160 mg of NaOH (4 mmol) in 10 g DEG. Inject the resulting solution into the three-neck flask using a syringe.
- **1.5** Stir and heat the solution to 220 °C for 2 h under nitrogen flow.
- **1.6** While heating, dissolve 153.17 mg of dopamine (1 mmol) in 0.4 ml of distilled water, and mix with 5 g of DEG.
- **1.7** Inject the dopamine solution into the reaction mixture at the end of heating.
- **1.8** After cooling the reaction mixture to room temperature, add 50-ml of ethanol to the three-neck flask.
- **1.9** Separate the dopamine-capped iron oxide nanoparticles by centrifugation at 10,000 rpm for 15 min.
- **1.10** Resuspend the particles in 50-ml ethanol using a sonicator bath.

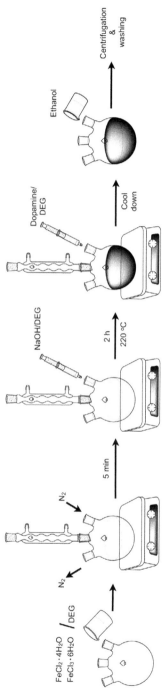

Figure 4 Step 1 schematic of nanoparticle synthesis.

1.11 Repeat the washing procedure with ethanol three times to remove excess chemicals.

1.12 To obtain pure dopamine-capped iron oxide nanoparticles, dry the particles under nitrogen flow.

1.13 Resuspend the particles in pH 5, 10 mM sodium acetate buffer for further use.

5.3 Tip

Dissolving NaOH in DEG will be difficult even with sonication. Make sure the NaOH powder is totally dissolved before using it.

5.4 Tip

The final concentration of the nanoparticles cannot be calculated by the weight of precursors because of potential loss of particles during the wash steps. To obtain the final concentration, weigh the dry powders after washing and quickly dissolve them in sodium acetate buffer to the required concentration. Avoid exposing the dry powder to air for extended periods of time.

5.5 Tip

To prevent oxidation of the particles, the drying process utilizes nitrogen atmosphere.

5.6 Tip

Nanoparticles dissolved in the buffer can be stored in a refrigerator for 3 months, and the nanoparticle powders should be kept in nitrogen-filled vials, or suspended in ethanol.

5.7 Tip

The procedure can be scaled up or down. Keep the molar ratio of precursors the same and adjust the volume of DEG to maintain the correct concentration.

6. STEP 2: PURIFICATION OF ENZYME

6.1 Overview

During this step, crude lipase will be purified prior to assembly of the microparticles. Lipase is purified by filtration and centrifugation to remove storage

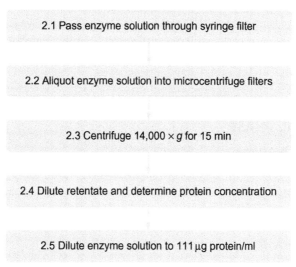

Figure 5 Flowchart of Step 2.

material (e.g., sugars, glycerol) which could negatively impact self-assembly and core synthesis (Fig. 5). Final protein concentration can be quantified using standard methods (e.g., bicinchoninic acid, Coomassie Blue) by comparison to a standard curve of bovine serum albumin.

6.2 Duration

30 min

2.1 Pass 2-ml of a 10% (w/v) solution of lipase in pH 7, 10 mM MES buffer through a 0.22-μm syringe filter attached to a 0.5-ml Luer-Lock syringe.
2.2 Distribute 400 μl aliquots of the enzyme solution into (2) 0.5-ml 10K MWCO microcentrifuge tubes (i.e., EMD Millipore Amicon Ultra 0.5).
2.3 Centrifuge at $14,000 \times g$ for 15 min.
2.4 Dilute retentate to approximately 1 mg/ml protein. Check protein concentration in a spectrophotometer for absorbance at 280 nm.
2.5 Dilute lipase in pH 7, 10 mM MES buffer to 111 μg protein/ml.

6.3 Tip

The amount of lipase prepared depends on the total number of planned samples. This protocol may be altered based on the amount of microparticles needed.

6.4 Tip

Use prepared lipase within 5 h of purification. Native lipase loses activity in solution over some time.

7. STEP 3: PREPARATION OF THE AQUEOUS PHASE AND OIL PHASE

7.1 Overview

Dopamine-capped iron oxide core nanoparticles and lipase are electrostatically associated prior to emulsification with an oil phase containing Grubbs' catalyst. Aliquots of the monomer are prepared independent of the oil phase and the catalyst to prevent polymerization of the monomer, prior to emulsification with the aqueous phase.

7.2 Duration

25 min

3.1 Pipette 450 µl lipase at a concentration of 111 µg protein/ml into (4) 1.5-ml microcentrifuge tubes.
3.2 Add 450 µl iron oxide nanoparticles at a concentration of 200 µg/ml into each 1.5-ml microcentrifuge tube containing lipase.
3.3 Rotate the enzyme–nanoparticle solutions on a mini rotator for 5 min at room temperature.
3.4 Pipette 90 µl dicyclopentadiene into (4) 1.5-ml microcentrifuge tubes and set aside.
3.5 Weigh 20 mg of Grubbs' catalyst (first generation) into a new 1.5-ml microcentrifuge tube.
3.6 Add 200 µl of toluene to the microcentrifuge tube containing Grubbs' catalyst.
3.7 Sonicate the Grubbs' catalyst and toluene for 10 min.
3.8 Pipette 150 µl of trichlorobenzene into a 1.5-ml microcentrifuge tube.
3.9 After complete dissolution, add 150 µl of the Grubbs' catalyst and toluene solution to the microcentrifuge tube containing the trichlorobenzene.
3.10 Mix well using a vortex.

7.3 Tip

Discharge static electricity before weighing Grubbs' catalyst to ensure accuracy.

7.4 Tip

Prepare oil phase ingredients in a chemical fume hood. The monomer, dicyclopentadiene, is volatile and has a strong odor.

7.5 Tip

Grubbs' catalyst can be difficult to dissolve. Additional time in the sonicator may be needed to fully solubilize the catalyst in toluene. When particulate matter is no longer visible in the toluene phase, the catalyst is fully dissolved.

7.6 Tip

Grubbs' catalyst is light sensitive. Avoid exposure to light as much as possible. While not in use, cover the microcentrifuge tube containing the catalyst with aluminum foil.

7.7 Tip

The polymerization happens as soon as the DCPD oil is mixed with the catalyst. To allow nanoparticle/enzyme to assemble on the oil/water interface, slow polymerization rate is necessary. Note that too low an amount of the catalyst will lead to insufficient polymerization, forming soft microparticles. Too much catalyst, on the other hand, will accelerate the polymerization reaction, resulting in less loading of the enzyme/nanoparticles on microparticles.

7.8 Tip

Before forming oil emulsions, nanoparticles and enzymes form a complex via electrostatic interactions. Therefore, the molar ratio of nanoparticle and the enzyme needs to be adjusted to achieve the balance of positive and negative charges in order to avoid repulsive forces on the droplets of the emulsion. By measuring the zeta potential of nanoparticle/enzyme complexes, optimized ratio of nanoparticle/enzyme can be obtained when the zeta potential is around zero. To measure the zeta potential, complexes with different ratios of nanoparticle/enzyme are suspended in 5 mM sodium acetate buffer (pH 5), and the potential is measured at 25 °C using a Zetasizer.

7.9 Tip

The concentration of nanoparticle/enzyme complex can affect the size of the emulsion droplets and the resulting microparticles. In general, using higher concentrations of the complex results in smaller emulsion drops, vice versa.

8. STEP 4: MICROPARTICLE ASSEMBLY

8.1 Overview

Biocatalytic complexes are formed through hierarchical interfacial assembly using amalgamation of the oil phase and aqueous phase. Until this step, the aqueous phase and the organic phase remain separate (Fig. 6).

8.2 Duration

2 min per tube

4.1 Mix the oil phase using a vortex.
4.2 Pipette 10 µl into one 1.5-ml microcentrifuge tube containing 90 µl dicyclopentadiene.
4.3 Gently vortex the tube and transfer the contents into a tube containing the aqueous phase.
4.4 Place the microcentrifuge tube into the amalgamator and shake for 30 s, 4200 cpm.
4.5 Remove the microcentrifuge tube from the amalgamator.
4.6 Repeat the process for the duration of the samples.
4.7 Place the tubes in a microcentrifuge tube rack and allow the samples to polymerize for 5 h at room temperature.

8.3 Tip

After the oil phase is added to the monomer solution, work quickly to prevent excessive polymerization prior to amalgamation. Grubbs' catalyst begins polymerization when once it is mixed with the monomer.

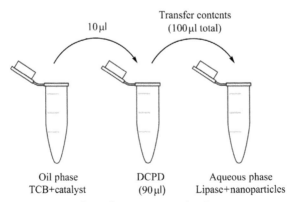

Figure 6 Step 4 process flow for microparticle formation, in preparation for amalgamation.

8.4 Tip

Avoid a rigorous vortex of the oil phase and monomer solution prior to addition with the aqueous phase. Vortex gently, just enough to disperse the monomer in the oil phase. The goal is to prevent the oil phase and monomer from adhering to the side of the microcentrifuge tube.

8.5 Tip

Increasing the shaking time for emulsification will result in smaller oil droplets; therefore, they would result in smaller polymer particles. Likewise, shortening the shaking time will form larger oil droplets with a wider size distribution. However, shaking for too long will generate heat during the emulsification and consequently it could denature the enzymes.

9. STEP 5: MICROPARTICLE WASHING

9.1 Overview

Microparticles are washed with nanopure water after polymerization is complete. This washing step removes residual reactants and prepares the sample for storage and future use. Microparticles are stored in buffer and retain activity at refrigeration temperatures.

9.2 Duration

30 min

5.1 Centrifuge the microcentrifuge tubes for 30 s at $2000 \times g$ (3.8 cm tube height).
5.2 Remove the supernatant from each microcentrifuge tube and dispose it off in a proper waste container.
5.3 Add 1 ml of nanopure water to each microcentrifuge tube and vortex.
5.4 Centrifuge the microcentrifuge tubes for 30 s at $2000 \times g$.
5.5 Repeat washing and centrifugation for a total of three washing cycles.
5.6 After the last round of microparticle washing, remove the supernatant and add 900 μl, pH 7, 10 mM MES buffer.
5.7 Vortex the microcentrifuge tubes to disperse the microparticles in the storage buffer.
5.8 Store the microparticles at 4 °C for further experiments.

9.3 Tip

Use caution not to discard of microparticles when removing the supernatant. If the supernatant in the pipette tip appears cloudy, microparticles have been removed.

9.4 Tip

Ensure microcentrifuge tubes are vortexed well. All particles should be suspended in solution before continuing to the next wash step. There should be no visible pellet at the bottom of the microcentrifuge tube.

10. CONCLUSIONS

The aforementioned protocol produces an average size particle of 10 μm (Talbert et al., 2014). Due to their size, microparticles can be easily removed from solution through centrifugation, magnetic separation, or gradual settling over time. Likewise, the microparticles can be resuspended through a simple vortex. Following interfacial assembly into hierarchical structures, the apparent catalytic efficiency increases to 1.9×10^4 M/s from 1.5×10^4 M/s, using resorufin butyrate as a substrate (Talbert et al., 2014). Additionally, these microparticles show no measurable protein leakage when exposed to extreme conditions (Talbert et al., 2014). This technology can be altered for different systems by tailoring properties such as enzyme type, nanoparticle capping ligand, ratio of nanoparticles to enzyme, as well as core material type. Lipase performs well in this system because of its enhanced activity when situated at a hydrophobic interface (Palomo et al., 2002).

REFERENCES

Adlercreutz, P. (2013). Immobilisation and application of lipases in organic media. *Chemical Society Reviews, 42*(15), 6406–6436.

Arroyo, M., Sánchez-Montero, J. M., & Sinisterra, J. V. (1999). Thermal stabilization of immobilized lipase B from Candida antarctica on different supports: Effect of water activity on enzymatic activity in organic media. *Enzyme and Microbial Technology, 24*(1–2), 3–12.

Bang, A., Mohite, D., Malik, A., & Nicholas, S. (2015). Polydicyclopentadiene aerogels from first- versus second-generation Grubbs' catalysts: A molecular versus a nanoscopic perspective. *Journal of Sol-Gel Science and Technology, 75*(2), 460–474.

Bastida, A., Sabuquillo, P., Armisen, P., Fernández-Lafuente, R., Huguet, J., & Guisán, J. M. (1998). A single step purification, immobilization, and hyperactivation of lipases via interfacial adsorption on strongly hydrophobic supports. *Biotechnology and Bioengineering, 58*(5), 486–493.

Dyal, A., Loos, K., Noto, M., Chang, S. W., Spagnoli, C., Shafi, K. V. P. M., ... Gross, R. A. (2003). Activity of Candida rugosa lipase immobilized on γ-Fe$_2$O$_3$ magnetic nanoparticles. *Journal of the American Chemical Society, 125*(7), 1684–1685.

Hasan, F., Shah, A. A., & Hameed, A. (2006). Industrial applications of microbial lipases. *Enzyme and Microbial Technology, 39*(2), 235–251.

Hwang, E. T., & Gu, M. B. (2013). Enzyme stabilization by nano/microsized hybrid materials. *Engineering in Life Science, 13*(1), 49–61.

Jeong, Y., Duncan, B., Park, M., Kim, C., & Rotello, V. M. (2011). Reusable biocatalytic crosslinked microparticles self-assembled from enzyme-nanoparticle complexes. *Chemical Communications, 47*(44), 12077–12079.

Jun, C., Jeon, B. W., Joo, J. C., Le, Q. A. T., Gu, S., Byun, S., ... Kim, Y. H. (2013). Thermostabilization of Candida antarctica lipase B by double immobilization: Adsorption on a macroporous polyacrylate carrier and R1 silaffin-mediated biosilicification. *Process Biochemistry, 48*(8), 1181–1187.

Palomo, J. M., Muñoz, G., Fernández-Lorente, G., Mateo, C., Fernández-Lafuente, R., & Guisán, J. M. (2002). Interfacial adsorption of lipases on very hydrophobic support (octadecyl–Sepabeads): Immobilization, hyperactivation and stabilization of the open form of lipases. *Journal of Molecular Catalysis B: Enzymatic, 19–20*, 279–286.

Poppe, J. K., Costa, A. P. O., Brasil, M. C., Rodrigues, R. C., & Ayub, M. A. Z. (2013). Multipoint covalent immobilization of lipases on aldehyde-activated support: Characterization and application in transesterification reaction. *Journal of Molecular Catalysis B: Enzymatic, 94*, 57–62.

Qu, H., Caruntu, D., Liu, H., & O'Connor, C. J. (2011). Water-dispersible iron oxide magnetic nanoparticles with versatile surface functionalities. *Langmuir, 27*(6), 2271–2278.

Samanta, B., Yang, X., Ofir, Y., Park, M., Patra, D., Agasti, S. S., ... Rotello, V. M. (2009). Catalytic microcapsules assembled from enzyme nanoparticle conjugates at oil water interfaces. *Angewandte Chemie International Edition, 48*(29), 5341–5344.

Sheldon, R. A. (2011). Characteristic features and biotechnological applications of cross-linked enzyme aggregates (CLEAs). *Applied Microbiology and Biotechnology, 92*(3), 467–477.

Sheldon, R. A., & Pelt, S. v. (2013). Enzyme immobilisation in biocatalysis: Why, what and how. *Chemical Society Reviews, 42*(15), 6223–6235.

Talbert, J. N., & Goddard, J. M. (2012). Enzymes on material surfaces. *Colloids and Surfaces B: Biointerfaces, 93*, 8–19.

Talbert, J. N., Wang, L., Duncan, B., Jeong, Y., Andler, S. M., Rotello, V. M., & Goddard, J. M. (2014). Immobilization and stabilization of lipase (CaLB) through hierarchical interfacial assembly. *Biomacromolecules, 15*(11), 3915–3922.

Wiemann, L., Nieguth, R., Eckstein, M., Naumann, M., Thum, O., & Ansorge-Schumacher, M. (2009). Composite particles of Novozyme 435 and silicone: Advancing technical applicability of macroporous enzyme carriers. *ChemCatChem, 1*(4), 455–462.

Yagonia, C. F. J., Park, K., & Yoo, Y. J. (2013). Immobilization of Candida antarctica lipase B on the surface of modified sol-gel matrix. *Journal of Sol-Gel Science and Technology, 69*(3), 564–570.

CHAPTER TWO

Monitoring Enzymatic Proteolysis Using Either Enzyme- or Substrate-Bioconjugated Quantum Dots

S.A. Díaz*,†, J.C. Breger*,†, I.L. Medintz*,1

*Center for Bio/Molecular Science and Engineering, Code 6900, U.S. Naval Research Laboratory, Washington, District of Columbia, USA
†American Society for Engineering Education, Washington, District of Columbia, USA
1Corresponding author: e-mail address: Igor.medintz@nrl.navy.mil

Contents

1. Introduction — 20
   1.1 Enzyme–Nanoparticle Constructs — 20
   1.2 Quantification Assay for Observing Modified Kinetics with Enzyme–QD Conjugates — 22
   1.3 Enzyme Activity Sensors Based on Transient QD–Enzyme Interactions — 23
2. Quantification Assay for Observing Modified Kinetics with Enzyme–QD Conjugates — 24
   2.1 Materials — 24
   2.2 Enzyme Assembly onto QDs — 25
   2.3 Obtaining Kinetic Data of QD-Enzyme Constructs — 29
   2.4 Analyzing the Data — 32
3. Enzyme Activity Sensors Based on Transient QD–Enzyme Interactions — 34
   3.1 Materials — 34
   3.2 Peptide Labeling — 36
   3.3 Peptide Precleaving — 39
   3.4 QD–Peptide Construct — 40
   3.5 Spectral Characterization — 43
   3.6 Fixed Enzyme Experiments — 45
   3.7 Fixed Substrate Experiments — 46
   3.8 Data Analysis — 47
4. Notes — 49
Acknowledgments — 50
References — 50

Abstract

Rational design of enzyme–nanoparticle hybrids is still in its infancy and the design is often inspired by potential access to many beneficial sensing properties such as increased stability, sensitivity, and even enhanced enzyme activities in specific cases. Deriving quantitative kinetic data from these constructs is not trivial, however, since the intrinsic design gives rise to unique properties that can influence the enzymatic assays that are central to the application of the hybrids. Here, we present two distinct assay methodologies for following the kinetic activity of composite enzyme–nanoparticle constructs. We utilize luminescent semiconductor nanocrystals or quantum dots (QDs) as the prototypical nanoparticulate platform for these sensing formats and target proteolytic enzyme activity as the main assay. The first assay is analogous to most current enzymatic assays and is designed to compare QD–enzyme constructs; this format is based on utilizing a fixed concentration of enzyme displayed on the QD and excess substrate in the solution, and the analysis utilizes data from initial velocities. The second assay is designed to analyze kinetics using a QD–substrate construct, in which the enzyme and QD interactions are short lived. Here, the nanoparticle–substrate concentration is held constant and exposed to increasing concentrations of the enzyme in solution. This later methodology is based on a fluorescent ratiometric signal that follows the entire progress curve of the enzyme reaction. A comparison of these two different assays of the series of enzyme–nanoparticle and substrate–nanoparticle constructs provides deeper insight into the enzyme kinetics of these hybrids, while still testing of individual variables within a given format, to allow for further optimization within each set.

1. INTRODUCTION
1.1 Enzyme–Nanoparticle Constructs

Rational modification of enzyme activity and enzyme kinetic parameters is valuable to a variety of practical applications. Along with classical studies on environmental optimization of enzymes for improved properties, bioengineering aims to improve activity, stability, and specificity of enzymes and it has widened the use of enzymes in both the industrial (Zaks, 2001) and academic worlds. Some advanced examples include: use in logic devices, hybrid systems, cell imaging, and redox sensors (Deshapriya et al., 2015; Dwyer et al., 2015; Hemmatian et al., 2015). The advent of controlled nanoparticle (NP) synthesis and NP functionalization has provided a new toolbox for increasing the available output from enzymes. NPs are defined by having spatial parameters in the nm range as opposed to the μm range for microparticles, while increasing the accessible surface area per unit mass of the particle; a wide variety of NPs are available today.

The current methodology, however, focuses on inorganic colloidal NPs with sizes typically less than 25 nm in diameter. Again a variety of materials are available such as iron oxide, noble metals (Au, Ag), and semiconductor nanocrystals or quantum dots (QDs; Sapsford et al., 2013). These particles, due to their nm size, present properties that differ from their bulk components including an increased surface/volume ratio, modified quantum-confined electronic, optical, and magnetic properties, as well as a capability to restructure their local environment in comparison to the bulk solution (Medintz, Uyeda, Goldman, & Mattoussi, 2005; Pfeiffer et al., 2014). These same properties are what provide NPs their capability to improve enzyme activity as well as create sensors that can follow enzyme activity under a variety of conditions (Johnson, Russ Algar, Malanoski, Ancona, & Medintz, 2014).

Here, we focus on QDs as they have exploitable optical properties such as high brightness, good photostability, with narrow and controllable emission peaks which are ideal for the rational design of sensors. Their large surface to volume ratios support extensive, straightforward conjugation strategies for attachment of multiple moieties (both enzymes and substrates) which is useful for both developing sensors and when seeking to enhance enzyme activities. Moreover, QDs tend to have narrow size distributions while displaying extremely robust physicochemical and optical properties. These attributes make QDs robust and dependable for use in assays.

There are two solution-based general architectures for QD–conjugates involving enzymes and these will be the focus of sections 2 and 3 presented within this chapter. The first combination is the strong binding of the enzyme to the QD surface, such as binding via metal ion complexation. This approach has been known to provide additional stability to the bound enzyme, and recent investigations have observed that enhanced kinetics are obtained when the supporting substrate is properly designed (Algar, Ancona, Malanoski, Susumu, & Medintz, 2012; Algar, Malanoski, et al., 2012; Algar, Malonoski, et al., 2012; Breger, Walper, et al., 2015; Brown et al., 2015; Claussen et al., 2013; Deshapriya & Kumar, 2013; Ding, Cargill, Medintz, & Claussen, 2015; Walper, Turner, & Medintz, 2015).

In a second architecture, the substrate is conjugated to the QD, while leaving the enzyme freely diffusing from solution to the particle and hence, an entirely different kinetic scheme is involved from that of the former architecture described above. The enzyme then binds to the NP through various reversible interactions such as specific substrate sensing, and nonspecific electrostatic, Van der Waals, and hydrophobic interactions, among others. These interactions result in both positive and negative effects on the enzyme's reaction kinetics and hence, a rational control of these interactions is of importance (Díaz, Breger, et al., 2015).

1.2 Quantification Assay for Observing Modified Kinetics with Enzyme–QD Conjugates

Enzyme interactions with surfaces enzyme encapsulation provide stabilization of the enzyme under adverse conditions such as high temperatures, extreme pH, and high ionic concentrations (Chaudhari & Kumar, 2000; Kim, Grate, & Wang, 2006; Mudhivarthi et al., 2012). Such design of enzyme binding allows for simpler enzyme recycling to reuse and reduce the cost of the biocatalyst, and it also allows for quick separation of the enzyme from the medium-containing substrate and products, downstream. One disadvantage encountered with this architecture is low efficiency of the enzyme, when the enzyme when the enzyme is bound to large surfaces. This result could be due to restricted diffusion of the substrate to the active site and release of the product into the solution. This was not the case when the enzyme is bound to layered zirconium(IV) phosphate nanosheets, which is presumably due to the direct exposure of these enzyme-bound nanosheets or clusters of nanosheets to the aqueous media (Chaudhari & Kumar, 2000; Mudhuvarthi, Bhambhani, & Kumar, 2007). Along these lines, the diffusional limitation can be overcome by utilizing small NPs or QDs as the conjugate scaffold (Johnson et al., 2014). The proper design of the QD–enzyme construct entails multiple parameters such as orientationally directed conjugation, solution medium optimization, along with QD optimization which can include size, shape, and ligand choice. Enhancing enzyme kinetics is of clear interest to maximize product creation (biosynthesis) or substrate degradation (bioremediation).

The mechanisms of enzyme catalytic enhancements at surfaces and macroparticles are still not fully understood, though evidence suggests that the increased diffusion of NPs, the higher surface curvature, modified microenvironment, extended hydration shell, favorable conformational changes, and increased local density of the enzyme (i.e., avidity) are possible explanations (Breger, Ancona, et al., 2015; Johnson et al., 2014; Lata et al., 2015; Chaudhari & Kumar, 2000). Likely a combination of these factors plays a role in enhancing or inhibiting the enzyme kinetics, as the case may be. The methodology presented in Section 2 provides an assay that permits the comparison of the respective kinetics of QD–enzyme constructs, where sequential variable modifications and rational design can optimize the functional constructs.

As an example, the enzyme phosphotriesterase (PTE, MW $\sim$37 kDa) was selected due to its tractability and possible use in bioremediation/biosafety applications (Tsai et al., 2012). The PTE family of enzymes are able to catalyze the hydrolysis of organophosphate ester compounds such as the insecticide paraoxon (bioremediation) as well as structurally similar nerve agents including sarin (biosafety; Tsai et al., 2012). The version of PTE utilized was first described by the Raushel lab and requires a divalent metal which interacts with the substrate in the catalytic step (Bigley & Raushel, 2013). The choice of paraoxon as a substrate for PTE results in a hydrolysis product (4-nitrophenol) which has an absorption band at 405 nm which is not present in the paraoxon (Breger, Ancona, et al., 2015).

1.3 Enzyme Activity Sensors Based on Transient QD–Enzyme Interactions

The complimentary architecture of QD-based enzyme constructs is the conjugation of the enzyme's substrate to the QD surface, which leads to temporary interaction of the enzymes in the solution with the QD. The standard application is for enzyme sensors, both for determining intrinsic enzyme activity and for monitoring moieties that act as inhibitors or synergistic factors of enzyme activity. The QD–substrate design allows for the creation of truly sensitive and robust reporters based on the Förster resonance energy transfer (FRET) mechanism (Nagy, Gemmill, Delehanty, Medintz, & Sapsford, 2014; Wu, Petryayeva, Medintz, & Algar, 2014). Here, the QD acts as a FRET donor and scaffold for displaying multiple FRET acceptors attached to the substrate. The design is applicable to portable, lab-on-a-chip style detectors, as well as to large high-throughput assays (Petryayeva & Algar, 2013). The QD-based FRET type of sensors are of particular interest as they specifically test enzyme activity as compared to other common methodologies that quantify the enzyme amount (immunostaining) or precursor (mRNA) (Knudsen, Jepsen, & Ho, 2013). Due to the inherent limitations in colloidal NP concentration ranges the methodology adopted in Section 2 is generally not applicable for monitoring enzyme activity. One has to pay attention to the fact that the substrate is QD bound and not in solution; hence, one needs to be cognizant about some diffusional limitations or restricted access to the substrate. In Section 3, we detail the experimental system utilizing enzymatic progress curves to properly determine the kinetics of enzymes using a QD–substrate conjugate sensor.

Proteases, such as the elastase (MW ∼25.9 kDa), which catalyze the hydrolysis of peptide bonds are presented here. Proteases are crucial for cellular function and have multiple commercial uses, such as in food technology and medical applications, among others (Craik, Page, & Madison, 2011; Shahidi & Janak Kamil, 2001). Proteases are particularly apt for FRET-based sensors as they can cleave the substrate and separate the FRET acceptor from the QD donor, considerably changing the fluorescence emission of both components. Elastase cleaves the peptide bond through a multistep process on the carboxyl side of small, hydrophobic amino acids, for example, glycine, alanine, and valine. The first step is the formation of an adsorption complex between elastase and the peptide, the next step is a nucleophilic attack and release of the C-terminus of the peptide. The remaining peptide is hydrolyzed in a deacylation step and this final step regenerates elastase (Dimicoli, Renaud, & Bieth, 1980).

2. QUANTIFICATION ASSAY FOR OBSERVING MODIFIED KINETICS WITH ENZYME–QD CONJUGATES

2.1 Materials

2.1.1 Equipment
1. Personal safety equipment: safety glasses, laboratory coat, latex or nitrile gloves, and fume hood.
2. Chemical waste receptacles.
3. Ultrapure water with a resistivity of 18.2 MΩ cm, for example, from a Milli-Q purification system (Millipore, Billerica, MA).
4. UV–visible spectrophotometer such as a Nanodrop 2000 Spectrophotometer (Thermo Scientific).
5. Gel imaging system with ultraviolet transillumination such as BioRad ChemiDoc™ XRS+ with Image Lab™ software.
6. Benchtop mini centrifuge.
7. Spectrophotometric multiwall plate reader such as Tecan Infinite® M1000 PRO (Tecan).
8. 1.5 and 2.0 mL microcentrifuge tubes.
9. 12-well pipette such as 5–50 µL Finnpipette™ Novus Electronic Multichannel Pipette.
10. Automatic pipettes 100 and 1000 µL volumes such as Eppendorf Xplorer® electronic pipettor.

2.1.2 Reagents
1. *N*-Cyclohexyl-2-aminoethanesulfonic acid (CHES) buffer (Sigma C2885): MW 207.29, 150 m*M*, pH 8.5.
2. Paraoxon (Supleco N12816): MW 275.20 g/mol, density (ρ) 1.274 g/mL.
3. 4-Nitrophenol (Sigma Aldrich): MW 139.11 /mol.

4. Phosphotriesterase-His$_6$ (see Note 1).
5. Water-soluble QDs (CdSe/ZnS core/shell) coated with dihydrolipoic acid (DHLA)-derived compact ligand 4 (DHLA-CL-4 or CL-4; Blanco-Canosa et al., 2014; Clapp, Goldman, & Mattoussi, 2006; Susumu et al., 2011) specifically we used in-house synthesized 525 nm green-emitting CdSe/ZnS QD with 4.2 ± 0.5 nm diameters resulting in ~ 60 nm^2 of surface per particle. The final concentration of the synthesized particles is typically in the high μM range; the working concentration of as prepared QDs used in assays is typically in the 5–10 μM range. Commercially available QDs such as ITK-COOH (Thermo-Fisher) can be used in a similar manner (Boeneman et al., 2013).
6. Gel electrophoresis apparatus.
7. Electrophoresis running buffer: 90 mM Tris-borate ethylenediaminetetraacetic acid (EDTA; 2 mM), pH 8.0–8.5 (1 × TBE buffer).
8. Agarose gel: 1.5% w/v of low EEO agarose solution in running buffer.

2.2 Enzyme Assembly onto QDs

Phosphotriesterase (PTE) was assembled onto QD surfaces by simply mixing the two components together at the desired stoichiometry as long as the final enzyme concentration is sufficiently high to yield progress curves with sharp linear regions containing more than three linear data points at the lowest substrate concentration tested. This assay can be modified for use with any commercially purchased or in-house prepared enzyme with a terminal His$_6$ tag, and other conjugation strategies are available and can be adapted to a similar methodology (Blanco-Canosa et al., 2014). The His$_6$ tag should be placed in such a way as not to interfere with the active site of the enzyme. Figure 1 schematically shows how the enzyme will attach to the QD surface, the chosen hydrophilic ligand structure, and the hydrolysis of paraoxon.

1. Determine the ratios of PTE to QD to be tested. The maximum number of enzymes that can be assembled to a QD can be computer modeled using a previously described method (Prasuhn, Deschamps, et al., 2010). These results can be confirmed by performing gel electrophoresis (see Fig. 2) or dynamic light scattering (DLS; Section 3.4.2 provides additional insight into these techniques).
2. For example, the range of ratios for 525 nm QD could be 0–16 PTE to QD. Calculate the total number of picomoles of PTE and the total number of picomoles of QD needed to achieve the desired ratio for

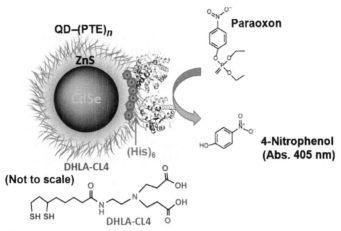

Figure 1 Schematic of enzyme–QD assembly. In this example, the enzyme is PTE and 525 nm CdSe/ZnS QDs have been rendered colloidally stable using the zwitterionic ligand DHLA-CL4. PTE is attached to the ZnS-rich surface of the QD through metal coordination with the His_6 engineered to the N-terminus end of the enzyme. PTE catalyzes the hydrolysis of paraoxon to 4-nitrophenol whose formation can be monitored by measuring the absorbance at 405 nm over time. *Adapted from Breger, Ancona, et al. (2015).* (See the color plate.)

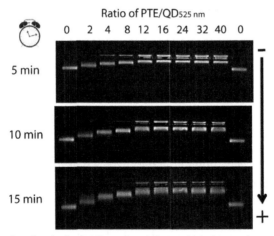

Figure 2 Example of gel electrophoresis performed on increasing amounts of PTE attached to QDs. The gel confirms assembly. Different ratios of PTE were attached to 525 nm CdSe/ZnS QDs ranging from 0 to 40. From computer modeling, theoretically the maximum number of PTE moieties that can be attached to 525 nm QDs is 10–16 (Breger, Ancona, et al., 2015). When run at 90 mV in running buffer, changes in electrophoretic mobility shifts in 1.5% w/v agarose gel confirm assembly of PTE to QDs at the various ratios. As the ratio of PTE to QD increases, the shift in electrophoretic mobility decreases. *Figure adapted from Breger, Ancona, et al. (2015).*

one condition (i.e., for a ratio of 2 PTE to QD, you would need 2 pmol of PTE and 1 pmol of QD; see Table 1). These should be prepared in labeled 1.5 mL centrifuge tubes (see Note 2).
3. Prepare equivalent samples of free enzyme. These contain no QD but are equivalent in enzyme concentration. These should be prepared at the same time as the PTE–QD samples in labeled 1.5 mL centrifuge tubes.
4. Prepare 0.7 mL of a 0.01 µM stock solution in CHES buffer, pH 8.5 of QDs in a 1.5 mL centrifuge tube.
5. Dilute concentrated PTE down to 0.1 µM in CHES buffer, pH 8.5 in a 1.5 mL centrifuge tube.
6. Determine the total volume of sample you will need for each condition by first determining the delivery volume for one well (i.e., 25 µL). Then multiply this volume by the total number of wells needed for each condition plus one or two extra wells (i.e., total number of wells = 18 + 2 extra wells = 20, 25 µL × 20 wells = 500 µL total reaction volume). The extra volume will account for any pipetting error (see Table 1).

Table 1 Ratios of PTE to QD Samples for Testing

	Ratio of Enzyme to QD						Enzyme Only					
	0	2	4	6	8	16	"0"	"2"	"4"	"6"	"8"	"16"
[QD] 0.01 µM (µL)	100	100	100	100	100	100	0	0	0	0	0	0
[PTE] 0.1 µM (µL)	0	20	40	60	80	160	0	20	40	60	80	160
Buffer (µL)	400	380	360	340	320	240	500	480	460	440	420	340
Total volume (µL)	500	500	500	500	500	500	500	500	500	500	500	500
[Final PTE] (nM)	0	4	8	12	16	32	0	4	8	12	16	32
[Final QD] (nM)	2	2	2	2	2	2	0	0	0	0	0	0
Picomoles of QD	1	1	1	1	1	1	0	0	0	0	0	0
Picomoles of PTE	0	2	4	6	8	16	0	2	4	6	8	16

This table is an example of how to make stock solutions of PTE–QD conjugates for a range of PTE to QD ratios. Stock solutions containing no QDs and only enzyme are also made at the same time. These stock solutions should contain enough volume to aliquot out for each substrate condition tested as well as replicates.

7. Select the QD concentration to be used, i.e., a concentration of 2 nM will provide sufficient number of QDs for PTE assembly, but the QD is diluted enough so as not to affect the absorbance measurements of the paraoxon hydrolysis.
8. In the following order add: buffer, enzyme, and QD to a microcentrifuge tube. For equivalent or free enzyme samples, do not add any QD but utilize the exact same buffer to make up the volume. Do this throughout for all similar control samples. Vortex briefly (<2 s) then briefly centrifuge with a mini centrifuge. Let stand at room temperature for ≥30 min. The samples are ready for use.
9. Aliquot 25 μL of sample to each well of every other column (each column corresponds to an enzyme sample) in a 384 white, clear bottom well plate (Corning 3706) using a 100 μL automatic pipette. When aliquoting, skip every other column. To aid the eye in pipetting, it is useful to mark every other column and every four rows with a black sharpie marker (see Fig. 3).
10. After adding samples, lightly tap the plate on the benchtop a few times to ensure the liquid falls to the bottom of well. No droplets should cling to the sides of the wells. If they fall during the measurements in the plate reader, the readings will be skewed.

Figure 3 Plate map for the addition of QD/enzyme or enzyme samples with markings to aid the eye for pipetting. Samples should be added to every other column.

2.3 Obtaining Kinetic Data of QD-Enzyme Constructs

1. Dissolve 3 µL of paraoxon (Supleco N12816-100MG) into 1 mL CHES buffer. As paraoxon is toxic, use full safety equipment and realize substrate preparation in a chemical hood. This should yield a 13.9 mM stock solution.
2. Perform a twofold serial dilution. A good range of concentrations is 5000–40 µM (in general, the range should attempt to cover 10-fold above and below the Michaelis constant; Cornish-Bowden, 2012) with a final volume around 800 µL for each concentration. The final concentration of substrate within the enzyme assay will range from 2500 to 20 µM (see Table 2).
3. Aliquot 60 µL of each substrate concentration to each well in a row of a 96-well plate (see Fig. 4) using a 1000 µL automatic pipette. It is recommended that the most concentrated substrate concentration be aliquoted into row A with each successive concentration going to each successive row. The least concentrated substrate will come to equilibrium first, therefore put it in last to obtain as many usable data points as possible, see Note 3.
4. Using a 12-well pipette, add 25 µL of each well in each row of the 96-well plate to the corresponding row of the 384-well plate. The total

Table 2 Substrate Dilution Table

Stock Solution =13.9 mM	Final [Substrate] (µM)	2500	1250	625	313	157	78	39	20
	[Substrate] (µM)	5000	2500	1250	625	313	157	78	39
	Volume of previous solution (µL)	576 (stock solution)	800	800	800	800	800	800	800
	Buffer volume (µL)	1024	800	800	800	800	800	800	800
	Total initial volume (µL)	1600	1600	1600	1600	1600	1600	1600	1600
	Final volume (µL)	800	800	800	800	800	800	800	1600

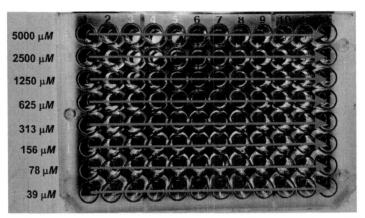

Figure 4 Plate map of 96-well plate containing substrate dilution samples. Add 60 μL of substrate at each concentration to each well of a row. Start with the most concentrated substrate sample.

volume in each well should be 50 μL. Repeat the same delivery scheme for the bottom half of the 384-well plate. (It is best to do this step right at the plate reader.) The tips of the 12-well pipette will naturally correspond to every other column of the 384-well pipette. There will be enough solution in each well of the 96-well plate to do a repeat on the bottom half of the 384-well plate. Substrate addition is a crucial step because as soon as the substrate is added the reaction will start. Therefore, make sure that there are no distractions around, nothing to interfere with arm movement, and are not working in a cramped place.

5. Immediately start measuring the absorbance using a plate reader capable of performing a kinetic script. The kinetic script has been written and optimized before the start of the assay. For example, a Tecan Infinite M1000 plate reader was used. The 384-well plate was placed in the opening of the plate reader, substrate was added, and the kinetic script immediately started. For the most dilute substrate progress curves, there will be very few linear data points. The temperature was set to room temperature or to an evaluated desired temperature (up to 45 °C, which was the limit of our plate reader). Select every other column. A kinetic script consisting of performing the same task every 25 s: shaking the plate at an amplitude of 2–4 for up to 5 s before measuring the absorbance at 405 or 348 nm for a total of 3 h. The 4-nitrophenol presents the protonated and deprotonated conformations with considerable variance in the extinction coefficient, therefore the absorbance

should be measured at a basic pH (Bowers, McComb, Christensen, & Schaffer, 1980). A work around this problem is to use the isosbestic point at 348 nm which makes the absorbance measurement buffer independent. When finished with the experiment both the 384-well plate and the 96-well plate should be discarded into the appropriate receptacles (solid chemical waste).

6. After the kinetic script is finished, construct a product calibration curve consisting of twofold serially diluted 4-nitrophenol dissolved in CHES buffer. A good range is from 0 to 1 mg/mL with a suggested final volume of 200 μL. Aliquot 50 μL of each sample to a 384-well plate such that each concentration is performed in triplicate. Measure the absorbance at 405 ($\varepsilon = 18.4 \,\mathrm{mM}^{-1}\mathrm{cm}^{-1}$) or 348 nm ($\varepsilon = 5.4 \,\mathrm{mM}^{-1}\mathrm{cm}^{-1}$) with the same plate reader and settings used for the kinetic assays. The addition of the QD can be obviated as the extremely low concentration results in negligible absorbance at the chosen wavelength (Breger, Ancona, et al., 2015).

7. Construct the calibration curve by averaging three absorbance values together and plotting them against their respective concentration values. Throw out any concentrations and absorbance values if their absorbance values exceed the detection limits of the plate reader.

8. Perform a linear regression analysis to obtain a linear equation to convert absorbance values to concentration of 4-nitrophenol (Fig. 5).

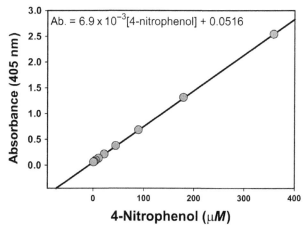

Figure 5 Calibration curve with resulting equation to convert absorbance values into product (4-nitrophenol) concentration values.

2.4 Analyzing the Data

1. Discard any data points that exceed the limits of the plate reader. This can be done by employing the IFERROR function in Excel™. (This function can be combined with the equation for converting absorbance values to concentration values, see Fig. 6.)
2. Initial enzymatic rates can be determined from the data in Excel™ either by using the SLOPE and RSQ functions on the linear portions of the progress curves (see Fig. 7) or by writing a macro to automate the calculations (highly recommended). Other software can also be used.
3. Once the initial rates are obtained, these can be plotted as a function of paraoxon substrate concentration and then be fitted to the Michaelis–Menten equation (see Eq. 1) either by using the enzyme kinetics module in SigmaPlot™ or by using the Solver add-in for Excel™ to fit the Michaelis–Menten curve by minimizing the sum of the error between the estimated initial rates and the actual rates using nonlinear least squares (see Fig. 8). The Michaelis–Menten equation is a mathematical model that relates substrate concentration to initial rate, assuming there is free diffusion of the substrate (Medintz et al., 2006). The reaction of enzyme and substrate can be schematically represented as Eq. (2). The k_{cat}

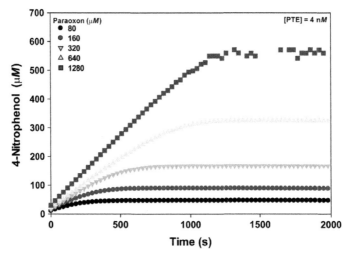

Figure 6 Typical progress curves obtained from the hydrolysis of paraoxon catalyzed by PTE. The absorbance values have been converted to product concentration values using a standard curve. Absorbance values that exceeded the detection limits of the plate reader were discarded. (See the color plate.)

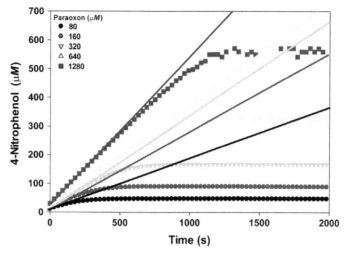

Figure 7 Representation of the linear fits utilized to determine the initial velocities from the data presented in Fig. 6. Only the initial linear portions of the progress curves for each substrate concentration should be utilized. These correspond to time windows of circa 0 to 150, 200, 250, 500, and 500 s for the 80, 160, 320, 640, and 1280 μM paraoxon substrate samples utilized, respectively, in this figure. (See the color plate.)

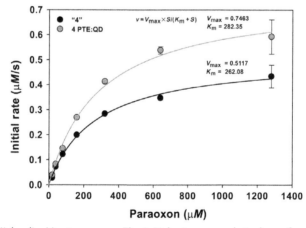

Figure 8 Michaelis–Menten curves. The initial rates were plotted as a function of substrate concentration for enzyme attached to QD (4 PTE:QD) and equivalent concentration of free enzyme. The data were fitted to the Michaelis–Menten equation to obtain V_{max} and K_M values.

parameter, or turnover rate, describes the maximum rate at which the enzyme (E) can transform substrate (S) into product (P). Within a specific experiment, V_{max} is the maximum achievable velocity; V_{max} depends on the k_{cat} value and the concentration of E in the solution.

The Michaelis constant, K_m, represents the substrate concentration at which the rate of the reaction is half of the maximum velocity. The K_m provides a good estimate for the enzyme's affinity for a substrate (a lower K_m signifies higher affinity, under specific conditions).

$$v = \frac{V_{max} \cdot S}{K_m + S}; V_{max} = k_{cat} \cdot E \qquad (1)$$

$$E + S \underset{k_r}{\overset{k_f}{\rightleftarrows}} ES \xrightarrow{k_{cat}} E + P \qquad (2)$$

3. ENZYME ACTIVITY SENSORS BASED ON TRANSIENT QD–ENZYME INTERACTIONS

3.1 Materials

3.1.1 General Equipment
1. Pipettes.
2. Vortex.
3. Speed–Vacuum system.
4. Syringes (1, 5, and 10 mL).
5. 1.5 and 2.0 mL microcentrifuge tubes.
6. Absorbance spectrophotometer.

3.1.2 Peptide Labeling
3.1.2.1 Equipment
1. Aluminum foil.
2. Eppendorf or tube rotator.
3. Snap-tight plastic cartridges with frits and Luer-Lok/slip fittings on each end, approximately 1.0 mL capacity.

3.1.2.2 Reagents
1. Commercial peptide with C-terminal cysteine residue and 6N-terminal Histidine (His$_6$-Tag) (see Note 4). As an example: H$_2$N-HHHHHHGGPPPPPPPAA(Aib)GAGAGAAPVSGC-COOH (see Fig. 9); Bio-Synthesis Inc., Aib is the artificial amino acid 2-aminoisobutyric acid.
2. Maleimide-modified organic dyes, such as Alexa Fluor 594 C$_5$ Maleimide (1 mg, ThermoFisher).
3. Succinimidyl ester-modified organic dyes, such as Alexa Fluor 647 NHS Ester Succinimidyl Ester (1 mg, ThermoFisher).

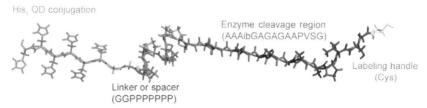

Figure 9 Schematic of elastase substrate peptide which binds to the QD. The four partitions mentioned in Section 3.2 are noted in the figure. (See the color plate.)

4. Ultrapure DNAase-free distilled H_2O.
5. Dimethyl sulfoxide (DMSO).
6. Phosphate-buffered saline (PBS) $10\times$ (100 mM phosphate, 1.37 M NaCl, 30 mM KCl, pH 7.4). Dilutions of PBS are prepared by diluting with the appropriate amount of Ultrapure H_2O.
7. Nickel (II)-nitrilotriacetic acid (Ni-NTA)-agarose (Qiagen).
8. 300 mM imidazole solution prepared in $1\times$ PBS.
9. RP18 oligopurification cartridge (Applied Biosystems).
10. Triethylamine acetate (TEAA) buffer 0.2 M, pH 7.
11. Acetonitrile (ACN); 70% ACN/H_2O solution is prepared with appropriate amount of Ultrapure H_2O.

3.1.3 Peptide Precleaving
3.1.3.1 Equipment
1. Temperature-controlled water baths or heating blocks.
2. Centrifuge/microcentrifuge.

3.1.3.2 Reagents
1. Dye-labeled peptide such as the Alexa Fluor 594-labeled His_6-Tag containing substrate peptide (Section 3.2.1).
2. Assay enzyme, in this case, Type III elastase from porcine pancreas (Sigma-Aldrich).
3. Enzyme active buffer, in the case of elastase, $1\times$ PBS + 10 µM $CaCl_2$ (ePBS) as Ca(II) is required by the enzyme.

3.1.4 QD–Peptide Construct
3.1.4.1 Equipment
1. Agarose gel electrophoresis.
2. Gel imaging system with ultraviolet transillumination.
3. Dynamic light scattering

3.1.4.2 Reagents
1. Water-soluble QDs such as 540 nm emitting CdSe/ZnS core/shell QDs with DHLA-Compact Ligand 4 (CL-4) zwitterionic surface ligands. See Fig. 1 for ligand structure.
2. Dye-labeled peptide (Section 3.2.1).
3. Enzyme active buffer, in the case of elastase: 1× PBS + 10 μM CaCl$_2$ (ePBS).
4. Electrophoresis running buffer: Tris/Borate/EDTA buffer (1× TBE, 100 mM Tris, 90 mM boric acid, 1 mM EDTA, pH 8.4).
5. DMSO.
6. Agarose gel (low EEO).

3.1.5 Spectral Characterization and Kinetics Measurements
3.1.5.1 Equipment
1. Spectrophotometric multiwall plate reader such as a Tecan Infinite® M1000 PRO (Tecan).
2. 96-well microtiter plates.
3. Automatic pipette capable of 50 μL volumes.

3.1.5.2 Reagents
1. Items 1–3 from Section 3.1.4.2.
2. Predigested peptide (see Section 3.3).
3. Dual-labeled peptide (See Section 3.2.2).

3.2 Peptide Labeling

The optimum peptide design for enzyme assays is composed of four sections. The first is the section for QD conjugation. Multiple options are available for QD conjugation (Algar et al., 2013; Blanco-Canosa et al., 2014; Sapsford et al., 2013); we will focus on using polyhistidine tags, specifically His$_6$, as it has been demonstrated to efficiently and ratiometrically conjugate to QDs in aqueous buffer in a simple mix and assemble strategy (Medintz et al., 2003; Prasuhn, Blanco-Canosa, et al., 2010). The second section is the linker or spacer section, again multiple options are available, and one of the most commonly utilized is the polyproline spacer as it provides a rigid rod-like structure whose length generally correlates with the number of proline repeats (Best et al., 2007; Gemmill et al., 2015). It is important to consider the size and mechanism of substrate recognition of your enzyme when designing the linker length. If the linker is of inadequate length for the enzyme to bind without being sterically hindered by the QD, your enzyme kinetics will be inhibited. If the linker is excessively long, then the energy transfer from the QD to the acceptor will be minimal, resulting in a small readout signal. The enzyme's substrate recognition motif is the third section and is included after the spacer region. The final section is the handle for dye labeling. Peptides can be commercially obtained with varying labeling handles, the simplest being the thiol group belonging to a cysteine (Kim et al., 2008). If Cys is not present in the rest of the peptide, then targeted labeling can be obtained by placing a Cys at the C-terminal section of the peptide. In general, using the N-terminal section for QD binding section (the His_6-tag) allows for subsequent labeling of the N-terminal amine if it becomes necessary to have a dual-labeled peptide. The example we will present contains a His_6-Tag for QD conjugation, 2 Gly followed by a Pro_7 spacer, the enzyme recognition and proteolysis motif, and a terminal cysteine.

3.2.1 Single Cysteine Labeling

1. Maleimide-modified dyes are commercially available as dried reagents and will react efficiently and specifically with thiol-reactive groups at neutral pH. The reagent is provided in an eppendorf tube containing 1 mg of Alexa Fluor 594 C_5 Maleimide (1 mg, 1.1 µmol, Thermo-Fisher). They can be dissolved in minimal amounts of DMSO (20 µL) and H_2O (80 µL) directly added to the dye eppendorf. Dye preparation should be realized right before conjugation.
2. Commercial peptides will generally be provided as lyophilized powders. Dissolve the peptide (1.0 mg of 2750 Da peptide, 0.35 µmol) in an adequate aqueous buffer (see Note 5). 350 µL of 10 × PBS is added to the elastase substrate peptide. The peptide is vortexed and allowed to stand for 5 min with occasional vortexing then added to the eppendorf containing the dye. A second addition of 350 µL of 10 × PBS can be added to the original peptide vessel and after 5 min added to the dye eppendorf. The dye is in a >threefold excess.
3. The sample is covered in aluminum foil and left in an eppendorf rotator at 30 rpm overnight.

3.2.2 Dual Labeling for Control Experiments

1. The dual-labeled peptides exploit the orthogonal labeling of thiol and amine groups. They are prepared as above with the inclusion of a similar molar excess of NHS ester succinimidyl ester-modified secondary dye. The selected dyes should form an efficient FRET donor acceptor pair (see Fig. 10), and we selected the Alexa Fluor 647 NHS Ester

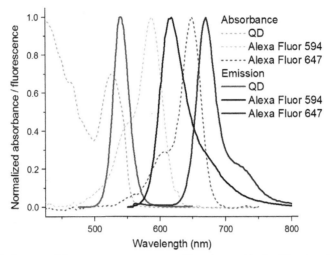

Figure 10 Spectra of fluorescent components. Absorbance (dashed lines) and Emission (full lines) are normalized to a maximum peak to demonstrate spectral overlap. (See the color plate.)

Succinimidyl Ester (1 mg, 0.8 μmol, ThermoFisher) as a complement to the Alexa Fluor 594.
2. It is not required that both ends be labeled at the same time. A subsequent dye labeling can be realized in the same manner as the first labeling. In this case, the cysteine-labeled peptide replaces the unlabeled peptide. The order of peptide labeling is generally of no consequence.

3.2.3 Peptide Purification, Quantification, and Storage

1. The selection of a His_6-Tag for QD conjugation provides the additional benefit of a simple and quick purification process following dye labeling. The His_6-Tag will bind to Ni-NTA affinity columns or reverse-phase cartridges allowing the peptides to be separated from unreacted dye. The strategy is applicable to single- or double-labeled peptides in equal measure. Labeling and purification is based on the specific protocol discussed in detail in the reference (Sapsford et al., 2009).
2. Utilize a Ni-NTA-agarose solution (~1.5 mL) as medium within a plastic cartridge (see Note 6). The medium must be washed of original solvent (EtOH) utilizing 10 mL of 10 × PBS equilibration buffer. It is estimated that 1 mL of medium will efficiently bind ~1 mg of His_6-Tag, if purifying more peptide prepare additional affinity cartridges.
3. Load cartridge with reaction solution utilizing a dual-syringe method (see Note 7) with 8–10 passes of the reaction through the cartridge.

The cartridge media should take the color of the reaction solution. If there is more peptide than can be bound by the cartridge, save the reaction medium and utilize a second cartridge, to bind additional peptide utilizing the same methodology.

4. Cartridge is then cleaned by 10 mL of 1× PBS, eliminating the unreacted dye reagent. Cartridge should remain colored as dye-labeled peptide is still bound.
5. The peptide is eluted by 1 mL additions of 300 mM imidazole in PBS (3×).
6. The peptide will require a desalting step to be utilized in subsequent assays. An RP18 oligopurification cartridge (Applied Biosystems) that was previously equilibrated with 5 mL of ACN and then 5 mL of 0.2 M TEAA buffer is the preferred choice for desalting.
7. Load RP18 cartridge with peptide solution utilizing the dual-syringe method with 8–10 passes of the peptide through the cartridge. The peptide solution should lose color as it passes through the resin of the cartridge, if RP18 cartridge saturates before solution decolorization utilize a second cartridge or regenerate cartridge (see Step 6) after obtaining the final peptide and reutilize.
8. The peptide-loaded RP18 cartridge is washed with 15 mL of 0.2 M TEAA buffer to remove contaminants.
9. Peptide is eluted from RP18 cartridge utilizing a 70% ACN/H_2O solution (3 × 0.5 mL). Measure absorbance spectra to determine final concentration (dilution may be necessary) utilizing the extinction coefficient values of the selected dye labels.
10. Aliquot samples into individual eppendorfs and dry in speed-vac system, store aliquots at −20 °C until use.

3.3 Peptide Precleaving

1. For proper calibration curves, a predigested peptide labeled with the cysteine reactive dye will be required (Algar, Malonoski, et al., 2012; Díaz, Malonoski, et al., 2015). Continuing the example, we will utilize the elastase enzyme and the corresponding elastase substrate peptide previously labeled with Alexa Fluor 594.
2. Resuspend the peptide (0.2 mg, 0.06 µmol) in ePBS (0.5 mL, 1 × PBS + 10 µM $CaCl_2$).
3. Add a relative excess of enzyme (0.05 mg, 1.9 nmol) to the peptide in an eppendorf. Incubate the reaction overnight at 37 °C to ensure complete digestion.

4. Place sealed Eppendorfs in boiling water bath for 1 h. This will denature the elastase enzyme (normal denaturing temperature is 58 °C; Favre-Bonvin, Bostancioglu, & Wallach, 1986) to ensure that there is no residual activity.
5. The boiled eppendorfs were centrifuged at 5000 rpm for 5 min to eliminate the enzyme pellet. Keep supernatant that contains digested peptides.
6. Though a ~100% yield is generally obtained, an absorbance spectra should be realized to obtain the precise final concentration. Aliquot the digested peptide and remove solvent in speed-vac system, store aliquots at −20 °C until needed.

3.4 QD–Peptide Construct
3.4.1 Formation

The His_6-Tag was selected for its ease of ratiometric assembly with QDs. This assembly through the formation of a metal complex is only stable in aqueous medium and any extreme conditions (i.e., extreme pH or ionic strengths) should be properly tested to assure peptide conjugation to the particles. As the conjugation is not a covalent bond, extremely dilute solutions may also affect the stability of the construct, and samples should be kept in the nM range. In this example, we will utilize water-soluble QDs with short zwitterionic surface ligands (compact ligand 4, CL-4, see Fig. 1; Susumu et al., 2011) In general, the peptide is able to bind to the QD even in the presence of bulkier surface moieties such as PEGylated ligands (Breger, Delehanty, & Medintz, 2015). The QD was selected to form a good FRET pair with the chosen organic label on the peptide, in this case a 540 nm emitting QD. In general, utilization of the QD as a donor is preferable to its use as a FRET acceptor (Algar, Wegner, et al., 2012; Petryayeva, Algar, & Medintz, 2013).
1. Dissolve the peptide in the desired buffer (1× PBS + 10 μM $CaCl_2$, ePBS) maintaining a concentration (1–10 μM) that is approximately an order of magnitude higher than the QD suspension. If the peptide is not particularly hydrophilic, a small amount (25 μL) of DMSO can be added before the buffer.
2. Determine the desired peptide/QD ratio (see Note 8).
3. Determine the precise QD concentration utilizing an absorbance spectra and the known extinction coefficient. In general, a concentration of 100–500 nM is recommended, if need be dilute sample with ePBS.

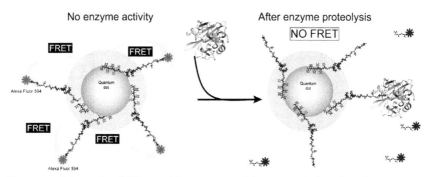

Figure 11 Schematic of QD–peptide enzyme activity sensor. Left: Before the presence of enzyme activity the substrate presents terminal FRET acceptors. Right: In the presence of enzyme activity the substrate is cleaved, the acceptor then diffuses away from the QD. The extended distance reduces the FRET efficiency, increasing QD emission, and decreasing the acceptor emission. (See the color plate.)

4. Calculate required volumes of peptide and QD solutions to obtain the desired ratio. Add the QD solution to the peptide in an eppendorf tube and allow 30 min for the conjugation to be completed (Fig. 11).

3.4.2 Characterization

The utilization of His_6-Tag containing peptides and water-soluble QDs has proven very robust (Blanco-Canosa et al., 2014; Petryayeva et al., 2013; Wu et al., 2014), yet, before continuing with the enzyme assays, it is important to ensure that the selected construct is not an exemption. This section deals with physical characterization with the next section detailing the spectroscopic experiments that can confirm QD–peptide construct assembly.

1. To characterize the efficiency and stability of the peptide–QD conjugation, agarose gels and DLS measurements can be realized.
2. For smaller QDs (6 nm in diameter, 110 nm^2 surface area), a maximum of 40–50 peptides can generally be bound (Prasuhn, Deschamps, et al., 2010). A series of different peptide/QD ratios should be prepared (recommended values could be 0, 1, 2, 4, 8, 12, 20, 30, 50). If possible, dilute all preparations to the same final concentration of QD (100 nM).
3. Prepare a 1% (w/v) agarose gel in TBE buffer with 10 wells each with 25 μL loading capacity.
4. Load 20 μL of QD–peptide solutions in the wells in increasing peptide/QD ratios starting and ending with the 0 peptide value.
5. Apply a voltage to the gel (150–200 V) and the QDs will move in the anode (+) direction. In general, as the peptide/QD ratio increases the

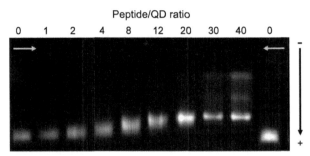

Figure 12 1% Agarose gel of varying peptide/QD ratio constructs; 20 minutes of run-time at 150 V. Horizontal arrows mark the initial sample loading positions. Starting at 30 and 40 peptide/QD ratios, we can observe the fluorescence from the unbound peptides.

mobility of the construct is decreased. After 10–30 min of run-time, a Gel-Imager system is utilized to observe the QD and peptide fluorescence (see Fig. 12). If it is possible to separate the fluorescence wavelengths emitted by the QD and the peptide, the colocalization of emission is an additional confirmation of conjugation. If excess peptide has been added, the band may be observed as a streak, precipitation may be observed, and/or there may be an additional band corresponding to unbound peptide. If the bands are not separated apply voltage for an additional 10 min of run-time to confirm the result.

6. Additionally, DLS can be used to observe the stability of the QD–peptide construct and to ensure that the constructs are not aggregating. The DLS measurements can give a rough estimate of QD–peptide conjugation efficiency, though they are less sensitive than gel electrophoresis (Choi, Pierson, Chang, Guo, & Kang, 2013).

7. The different peptide/QD ratios should be diluted with buffer (the DLS measurements should result in ~200 counts per second) and filtered with 0.2 μm syringe filters before any measurements (Oh et al., 2013).

8. The hydrodynamic radius determined by DLS will be larger than the physical radius determined for the QD by TEM. If the peptide is not excessively large, an approximately two-fold increase can be expected (Rameshwar et al., 2006). As the peptide/QD ratio increases, the observed hydrodynamic radius should increase, and the value saturates when the QD is fully covered in peptide (see Note 9).

3.5 Spectral Characterization

The spectral characterization should also confirm the proper conjugation of the peptide to the QD surface. As the QD and Alexa Fluor 594 are a FRET pair (Förster distance or $R_0 = 8.5$ nm, see Note 10), if properly conjugated, the QD emission will be quenched and there will be a sensitized signal or increased emission from the dye. An understanding of FRET theory, particularly in the context of a single donor (QD)–multiple acceptors (Alexa Fluor 594) system, is convenient but lies outside of the scope of this methodology. The following texts can be instructional for understanding FRET (Jares-Erijman & Jovin, 2003; Lakowicz, 2006; Medintz & Hildebrandt, 2014).

1. The absorbance spectra of both the QD and peptide should be obtained to determine the optimal wavelength for fluorescence excitation that maximizes QD emission and minimizes direct dye excitation.
2. We add 100 μL of our sample to a well within the 96-well plate.
3. The fluorescence spectra should be measured on the Tecan plate reader or another suitable instrument, and hardware conditions (slits, integration time, gain, etc.) optimized for the QD and the peptide individually at the range of concentrations that will be used in the latter plate reader assays. This signifies approximately 150 nM of QD and 150–7500 nM of peptide. In our example, the imaging conditions were: 5 nm slits, excitation 450 nm, 400 Hz flash frequency, detection at 540 and 617 nm when utilizing fixed wavelength, and 1 ns steps and 30 μs integration time for full spectral detection.

To obtain the calibration curves that will be required to transform the optical signal to an enzymatic assay, the following must be done.

1. A series of varying peptide/QD ratios should be prepared at the final concentration (150 nM of QD) of the assay using the ePBS buffer. The ratios should cover a wide range, i.e., 1–25 peptides/QD.
2. Obtain measurements of the full spectra as well as the fixed wavelength measurement format for all the constructs prepared above. If proper conjugation has occurred, then the QD emission should decrease and peptide label emission increase as the peptide/QD ratio increases. The change in signal saturates at some point.
3. Obtain the same measurements for predigested peptide/QD constructs covering the same range of peptide/QD ratios.
4. Plot both sets of data as emission ratio as a function of peptide/QD ratio (Fig. 13C). There should be a range of ratios at which the change in

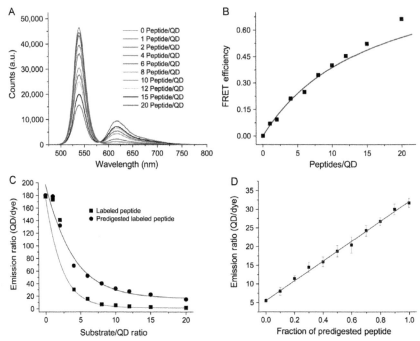

Figure 13 Calibration curves of enzyme sensor. (A) Fluorescence spectra of sensor as the number of peptide/QD increases. (B) FRET efficiency of the sensor as a function of the peptide/QD ratio. Squares are experimental data while the line corresponds to a FRET fit. It is important to note that at higher ratios deviation from the FRET formalisms begins to increase. (C) Emission ratios as a function of the substrate/QD ratio. It is important to note that a fully cleaved system is not equivalent to a system that has no dye in solution. Similarly, the curves demonstrate that at low substrate/QD ratios, the system is too volatile for confident measurements. The lines are simply to guide the eye, though proper numerical fits can be achieved. (D) Emission ratio of a 10 substrate/QD construct as a function of the fraction of predigested peptide. (See the color plate.)

signal is robust and consistent. This is the preferable range in which to realize the assays. This will in general eliminate work at smaller peptide/QD ratios (<5) as the signal can be excessively variant, as well as discard higher peptide/QD ratios (>20) as larger deviations from FRET theory begin to appear (internal filter effects and excessive direct excitation, see Note 11).

5. The final calibration curve is prepared by selecting a few peptide/QD ratios and testing the emission ratio as a function of the degree of digested peptide. As an example, 10 peptide/QD ratio can be tested and the fluorescence spectra obtained of the following combination:

10 peptide/0 predigested, 7.5 peptide/2.5 predigested, 5 peptide/5 predigested, 2.5 peptide/7.5 predigested, and 0 peptide/10 predigested. In general, a linear fit of the emission ratio as a function of the digested fraction is expected (Wu & Algar, 2015).

3.6 Fixed Enzyme Experiments

Fixed enzyme experiments with QD–Substrate constructs can be realized in the case of particularly efficient Michaelis Constant, K_m (low nM range), for the selected enzyme and peptide. It is also often used to determine the comparative kinetic parameters of the enzyme and peptide when the peptide is not conjugated to a QD. An example is presented in Section 2. We will specifically present the methodology for freely diffusing dual-labeled peptide but it can be adapted for any reagent.

1. The dual-labeled FRET competent substrate peptide (see Section 3.2.2) will be utilized for these experiments. A range of concentrations that reaches 10-fold excess of the enzyme–substrate K_m is preferable (in our example, 0–200 μM).
2. Pipette 100 μL of the maximum concentration substrate in a well within the 96-well plate and optimize the plate reader conditions (slits, gain, etc., see Note 12).
3. Using the 96-well plate create a line of serial dilutions of the dual-labeled substrate utilizing the ePBS buffer. Select a line of the 96-well plate and fill wells 2–12 with 50 μL of ePBS. In well 1, place 100 μL of a peptide solution that is twice as concentrated as the solution in Step 2 of Section 3.6. Take 50 μL of the solution from well 1 and add it to well 2. Mix the solution by pipetting up and down 3–4×. Repeat the process down the line, taking 50 μL of well X and adding it to $X+1$; the 50 μL from well 12 can be discarded. Every well in the line will contain 50 μL of final solution with decreasing substrate concentration (twofold dilution per well).
4. Prepare a dilution of the enzyme in ePBS with 700 μL final volume (200 pM elastase). Utilizing an automatic pipette add 50 μL to each well along the line. As soon as the additions are finished begin the fluorescence measurement script. The precise script will depend on your enzyme and its kinetic characterization, but it is important to have a good description of the initial velocity of the reaction, therefore taking measurements often in the initial stages. We recommend that some sort of mixing (∼2 s) be realized before the measurements begin.

5. Allow the reaction to reach completion, or if this is not feasible due to time constraints or turnover limitations realize a separate calibration. This is done by using the same peptide concentrations and adding a relative excess of enzyme to determine the reaction endpoint. Similarly a control with no enzyme, only ePBS buffer addition, should be realized to determine the initial emission ratio.

3.7 Fixed Substrate Experiments

One of the limitations of NP based enzyme assays is that the working concentrations are limited. The colloidal particles are generally limited to μM concentrations (Wu et al., 2014). If the sample becomes more concentrated then aggregation is observed as well as changes in the overall physical properties of the solution (i.e., viscosity and ionic strength). In the fixed enzyme kinetics methodology (see Sections 2.4 and 3.6), the substrate concentration must be added in large excess to obtain proper fittings (Cornish-Bowden, 2012). In general, this excess substrate concentration is above that achievable by colloidal NPs. To overcome this limitations, enzymatic progress curves are realized in a fixed substrate and varied enzyme concentration modality.

1. Determine the peptide/QD ratio that will compose your fixed substrate concentration (in our example, we will utilize 10 peptide/QD) and prepare 700 mL of the solution at $2\times$ the desired final concentration (solution concentration $=300$ nM of QD).
2. Prepare plate reader by introducing measurement script and allowing the system to reach the desired temperature (5 nm slits, excitation 450 nm, 400 Hz flash frequency, detection at 540 and 617 nm, at 25 °C for ~4 h measuring every 150 s).
3. Determine maximum enzyme concentration (600 nM) that you wish to study and dilute your enzyme in ePBS to twofold concentration (1200 nM). Select a line of the 96-well plate and fill wells 2–12 with 50 μL of ePBS. Add 100 μL of your 1200 nM enzyme solution to well 1; take 50 μL of the solution from well 1 and add it to well 2. Mix the solution by pipetting up and down 3–4 ×. Repeat the process down the line, taking 50 μL of well X and adding it to $X+1$ until reaching well 11. Well 12 should remain as only 50 μL of ePBS with no enzyme. Every well in the row will contain 50 μL of final solution with decreasing enzyme concentration (two-fold dilution per well) and a control "no enzyme" well on the end.

4. Place well plate in plate reader. Prepare an automatic pipette to complete 12 additions of 50 μL. Using the peptide/QD construct prepared in step 1 add 50 μL to each well along the row. As soon as the additions are finished begin the fluorescence measurement script. The precise script will depend on your enzyme and its kinetic characterization. We recommend that some sort of mixing (~2 s) be realized before the measurements begin.
5. In general, additional experiments (as well as duplicates) can be realized at the same time, though as each additional line will increase the lag time between addition of the enzyme and the beginning of the progress curve, we do not recommend realizing more than four rows at a time.

3.8 Data Analysis

The analysis of the fixed enzyme methodology is presented in Section 2 (see Figs. 6–8). In this section, we will present the fixed substrate methodology. As opposed to the fixed enzyme case where the initial velocity is sufficient, the entire progress curve is utilized for the parametric determination.

1. The raw emission ratio curves are obtained from the plate reader for each well. The curves are corrected for any experimental drift by normalizing to the control well containing no enzyme.
2. The emission ratio curves are transformed into peptide/QD curves utilizing the previously obtained calibration curves. As the QD concentration is known for all the wells, these curves can be represented as substrate concentration as a function of time.
3. The subsequent step is to transform these progress curves into *Enzyme Time* (Et) progress curves. This is done by multiplying the time progression by the enzyme concentration of each well, this results in an abscissa with typical units of ($\mu M \times s$) while the ordinate remains substrate concentration.
4. When the data are transformed and if the system can be represented by the MM parameters than the data should collapse into a large superimposed series of curves (see Fig. 14D). If the system is more complex and cannot be properly modeled by MM parameters, the specifics of the analysis will depend on each system and a common methodology cannot be described.
5. We will assume that the MM model can be applied to the data analysis as has been observed in various reported cases (Algar, Malonoski, et al., 2012; Breger et al., 2014; Claussen et al., 2015; Zhu et al., 2015).

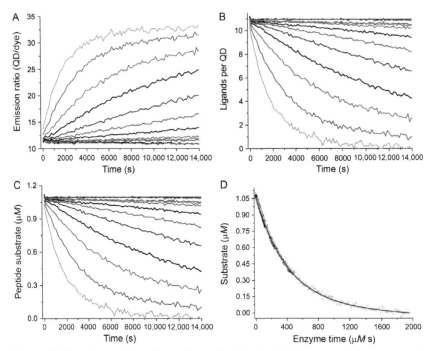

Figure 14 Step-by-step transformations of a fixed substrate enzyme sensor experiments. (A) Raw emission ratio curves. (B) Using the calibration curves and baseline correction to transform emission ratios into remaining ligands per QD. (C) Multiplication by QD concentration to obtain peptide substrate consumption progress curves. (D) Enzyme time peptide substrate consumption progress curves. The points are the overlying data points from the curves in part C. The blue line is the corresponding fit from Eq. (3). (See the color plate.)

6. The data points are combined and analyzed by using the integrated solution to the MM equations:

$$k_{cat} \times E_t = p + K_m \ln\left(\frac{s_0}{s_0 - p}\right) \quad (3)$$

7. Where k_{cat} and K_m are the *catalytic constant* and *Michaelis constant*, respectively. The reaction product is represented as p and the initial substrate as s_0, and E_t is enzyme time.
8. In general, individual k_{cat} and K_m values are not accessible, but their ratio, known as the specificity constant, can be obtained (see Note 13).

4. NOTES

1. PTE was isolated from *Escherichia coli* as described elsewhere (Susumu et al., 2014). Briefly, the gene encoding phosphotriesterase (EC 3.1.8.1) from *Brevundimonas diminuta* was synthesized by Genscript and cloned into a bacterial expression vector. The *E. coli* strain BL21 (DE3) strain was used for enzyme expression which was conducted at 30 °C for 24 h in 500 mL shake flasks. Enzyme was isolated from cell lysates using immobilized metal affinity chromatography and eluents were further separated via FPLC. Protein concentration was determined using a Nanodrop 1000 to measure absorbance at 280 nm and a calculation based on the extinction coefficient and molecular weight of the protein. The enzyme was diluted in 50–70% glycerol, aliquoted, snap frozen, and stored at −80 °C until use.
2. The total picomoles needed for the enzyme will depend on the overall activity of the enzyme, therefore some preliminary experimentation should be done with various concentrations of enzyme against varying substrate concentrations to determine the optimal final enzyme concentration needed for the assays.
3. The use of the 96-well plate is to optimize the subsequent loading and utilization of the multiarmed pipette, other setups can be used in replacement.
4. Peptides can be synthesized in-house; for simplicity, we will deal with purchased peptides.
5. Commercial peptides are generally provided with trifluoroacetate salts as counterions. A high molarity buffer is therefore generally recommended to ensure the proper pH.
6. Plastic cartridges are recuperated from spent OPC by snapping open the cartridge and eliminating the resin with multiple washes.
7. Briefly our recommended dual-syringe methodology is as follows. Select two 5 mL plastic syringes and remove the plunger from one and connect them both to either end of the Luer Fitting of a plastic cartridge. Place the system in a vertical holder with the syringe missing the plunger on top and load the reaction mixture in the open syringe. Make sure the syringe underneath is in the blocked position before replacing the plunger to the syringe above. You can then use the two syringes to push the reaction mixture through the cartridge multiple times.

8. An estimate for the maximum number of His$_6$-peptides that can be loaded on a spherical NP surface is 1 peptide for every 2.2 nm^2 of surface area.
9. DLS can be very sensitive. Larger particles may simply be dust or other impurities and not necessarily sample aggregation. If you suspect that aggregation is occurring realize DLS measurements at different time points.
10. R_0 was calculated using the equation $R_0^6 = C_0 \kappa^2 J n^{-4} Q_D$ where the QD quantum yield $(Q_D) = 0.2$, dipole orientation factor $(\kappa^2) = 2/3$, the refractive index $(n) = 1.33$, and J is the spectral overlap of the donor emission and the acceptor absorbance (Lakowicz, 2006). The $\kappa^2 = 2/3$ value has been shown to be appropriate for such randomly assembled QD constructs (Clapp et al., 2004; Medintz et al., 2004).
11. If precise calibration values are obtained, the deviation from FRET theory may not be an impediment to designing a QD-based enzyme sensor, but it is generally not recommended.
12. Wavelength detection should be at peak of both emitters and it is important that the donor emission be less than 1/3 the saturation value as this peak may increase considerably as the reaction proceeds.
13. Unless the following criteria can be met: (1) $a_0 > 3K_m$ and (2) Selwyn's test (Duggleby, 2001). Selwyn's test states that progress curves obtained from an MM system, at a fixed a_0 and variable E, will superimpose when time is scaled by E to yield the abscissa of enzyme-time. See Fig. 14 for an example of a dataset that meets Selwyn's test.

ACKNOWLEDGMENTS

S.A.D. and J.C.B. acknowledge the American Society for Engineering Education postdoctoral fellowship through the Naval Research Laboratory (NRL). I.L.M. acknowledges the office of Naval Research (ONR), NRL, NRL Institute for Nanoscience (NSI), and the Defense Threat Reduction Agency Joint Science and Technological Office (DTRA JSTO) MIPR # B112582M.

REFERENCES

Algar, W. R., Ancona, M. G., Malanoski, A. P., Susumu, K., & Medintz, I. L. (2012). Assembly of a concentric Förster resonance energy transfer relay on a quantum dot scaffold: Characterization and application to multiplexed protease sensing. *ACS Nano, 6,* 11044–11058.
Algar, W. R., Blanco-Canosa, J. B., Manthe, R. L., Susumu, K., Stewart, M. H., Dawson, P. E., et al. (2013). Synthesizing and modifying peptides for chemoselective ligation and assembly into quantum dot-peptide bioconjugates. *Methods in Molecular Biology, 1025,* 47–73.

Algar, W. R., Malanoski, A. P., Susumu, K., Stewart, M. H., Hildebrandt, N., & Medintz, I. L. (2012). Multiplexed tracking of coupled protease activity using a single color of quantum dot vector and a time-gated Förster resonance energy transfer relay. *Analytical Chemistry, 84*, 10136–10146.

Algar, W. R., Malanoski, A., Deschamps, J. R., Blanco-Canosa, J. B., Susumu, K., Stewart, M. H., et al. (2012). Proteolytic activity at quantum dot-conjugates: Kinetic analysis reveals enhanced enzyme activity and localized interfacial "hopping" *Nano Letters, 12*, 3793–3802.

Algar, W. R., Wegner, D., Huston, A. L., Blanco-Canosa, J. B., Stewart, M. H., Armstrong, A., et al. (2012). Quantum dots as simultaneous acceptors and donors in time-gated Förster resonance energy transfer relays: Characterization and biosensing. *Journal of the American Chemical Society, 134*, 1876–1891.

Best, R. B., Merchant, K. A., Gopich, I. V., Schuler, B., Bax, A., & Eaton, W. A. (2007). Effect of flexibility and cis residues in single-molecule FRET studies of polyproline. *Proceedings of the National Academy of Sciences of the United States of America, 104*, 18964–18969.

Bigley, A. N., & Raushel, F. M. (2013). Catalytic mechanisms for phosphotriesterases. *Biochimica et Biophysica Acta (BBA)—Proteins and Proteomics, 1834*, 443–453.

Blanco-Canosa, J. B., Wu, M., Susumu, K., Petryayeva, E., Jennings, T. L., Dawson, P. E., et al. (2014). Recent progress in the bioconjugation of quantum dots. *Coordination Chemistry Reviews, 263–264*, 101–137.

Boeneman, K., Delehanty, J. B., Blanco-Canosa, J. B., Susumu, K., Stewart, M. H., Oh, E., et al. (2013). Selecting improved peptidyl motifs for cytosolic delivery of disparate protein and nanoparticle materials. *ACS Nano, 7*, 3778–3796.

Bowers, G. N., Jr., McComb, R. B., Christensen, R. G., & Schaffer, R. (1980). High-purity 4-nitrophenol: Purification, characterization, and specifications for use as a spectrophotometric reference material. *Clinical Chemistry, 26*, 724–729.

Breger, J. C., Ancona, M. G., Walper, S. A., Oh, E., Susumu, K., Stewart, M. H., et al. (2015). Understanding how nanoparticle attachment enhances phosphotriesterase kinetic efficiency. *ACS Nano, 9*, 8491–8503.

Breger, J., Delehanty, J. B., & Medintz, I. L. (2015). Continuing progress toward controlled intracellular delivery of semiconductor quantum dots. *Wiley Interdisciplinary Reviews. Nanomedicine and Nanobiotechnology, 7*, 131–151.

Breger, J. C., Sapsford, K. E., Ganek, J., Susumu, K., Stewart, M. H., & Medintz, I. L. (2014). Detecting kallikrein proteolytic activity with peptide-quantum dot nanosensors. *ACS Applied Materials & Interfaces, 6*, 11529–11535.

Breger, J. C., Walper, S. A., Oh, E., Susumu, K., Stewart, M. H., Deschamps, J. R., et al. (2015). Quantum dot display enhances activity of a phosphotriesterase trimer. *Chemical Communications, 51*, 6403–6406.

Brown, C. W., III, Oh, E., Hastman, D. A., Jr., Walper, S. A., Susumu, K., Stewart, M. H., et al. (2015). Kinetic enhancement of the diffusion-limited enzyme beta-galactosidase when displayed with quantum dots. *RSC Advances, 5*, 93089–93094.

Chaudhari, A., & Kumar, C. V. (2000). Proteins immobilized at the galleries of layered α-zirconium phosphate: Structure and activity studies. *Journal of the American Chemical Society, 122*, 830–837.

Choi, M. J., Pierson, R., Chang, Y., Guo, H., & Kang, I. K. (2013). Enhanced intracellular uptake of CdTe quantum dots by conjugation of oligopeptides. *Journal of Nanomaterials. 2013*. Article ID 291020, 8 pages, http://dxdoi.org/10.1155/2013/291020.

Clapp, A. R., Goldman, E. R., & Mattoussi, H. (2006). Capping of CdSe-ZnS quantum dots with DHLA and subsequent conjugation with proteins. *Nature Protocols, 1*, 1258–1266.

Clapp, A. R., Medintz, I. L., Mauro, J. M., Fisher, B., Bawendi, M. G., & Mattoussi, H. (2004). Fluorescence resonance energy transfer between quantum dot donors and dye labeled protein acceptors. *Journal of the American Chemical Society, 126*, 301–310.

Claussen, J. C., Algar, W. R., Hildebrandt, N., Susumu, K., Ancona, M. G., & Medintz, I. L. (2013). Biophotonic logic devices based on quantum dots and temporally-staggered Förster energy transfer relays. *Nanoscale, 5*, 12156–12170.

Claussen, J. C., Malanoski, A., Breger, J. C., Oh, E., Walper, S. A., Susumu, K., et al. (2015). Probing the enzymatic activity of alkaline phosphatase within quantum dot bioconjugates. *Journal of Physical Chemistry C, 119*, 2208–2221.

Cornish-Bowden, A. (2012). *Fundamentals of enzyme kinetics* (4th ed.). Weinheim, Germany: Wiley-Blackwell.

Craik, C. S., Page, M. J., & Madison, E. L. (2011). Proteases as therapeutics. *Biochemical Journal, 435*, 1–16.

Deshapriya, I. K., & Kumar, C. V. (2013). Nano-bio interfaces: Charge control of enzyme/inorganic interfaces for advanced biocatalysis. *Langmuir, 29*, 14001–14016.

Deshapriya, I., Stromer, B. S., Kim, C. S., Patel, V., Gutkind, J. S., & Kumar, C. V. (2015). Novel, protein-based nanoparticles: Improved half-lives, retention of protein structure and activities. *Bioconjugate Chemistry, 26*, 396–404.

Díaz, S. A., Breger, J. C., Malanoski, A., Claussen, J. C., Walper, S. A., Ancona, M. G., et al. (2015). Modified kinetics of enzymes interacting with nanoparticles. In *Paper presented at the SPIE Nanoscience + Engineering, San Diego, USA*.

Díaz, S. A., Malonoski, A. P., Susumu, K., Hofele, R. V., Oh, E., & Medintz, I. L. (2015). Probing the kinetics of quantum dot-based proteolytic sensors. *Analytical and Bioanalytical Chemistry, 407*, 7307–7318.

Dimicoli, J.-L., Renaud, A., & Bieth, J. (1980). The indirect mechanism of action of the trifluoroacetyl peptides on elastase. *European Journal of Biochemistry, 107*, 423–432.

Ding, S., Cargill, A. A., Medintz, I. L., & Claussen, J. C. (2015). Increasing the activity of immobilized enzymes with nanoparticle conjugation. *Current Opinion in Biotechnology, 34*, 242–250.

Duggleby, R. G. (2001). Quantitative analysis of the time courses of enzyme-catalyzed reactions. *Methods, 24*, 168–174.

Dwyer, C. L., Díaz, S. A., Walper, S. A., Samanta, A., Susumu, K., Oh, E., et al. (2015). Chemoenzymatic sensitization of DNA photonic wires mediated through quantum dot energy transfer relays. *Chemistry of Materials, 27*, 6490–6494.

Favre-Bonvin, G., Bostancioglu, K., & Wallach, J. M. (1986). Ca2+ and Mg2+ protection against thermal denaturation of pancreatic elastase. *Biochemistry International, 13*, 983–989.

Gemmill, K. B., Díaz, S. A., Blanco-Canosa, J. B., Deschamps, J. R., Pons, T., Liu, H.-W., et al. (2015). Examining the polyproline nanoscopic ruler in the context of quantum dots. *Chemistry of Materials, 27*, 6222–6237.

Hemmatian, Z., Miyake, T., Deng, Y., Josberger, E. E., Keene, S., Kautz, R., et al. (2015). Taking electrons out of bioelectronics: Bioprotonic memories, transistors, and enzyme logic. *Journal of Materials Chemistry C, 3*, 6407–6512.

Jares-Erijman, E. A., & Jovin, T. M. (2003). FRET imaging. *Nature Biotechnology, 21*, 1387–1395.

Johnson, B. J., Algar, W. R., Malanoski, A. P., Ancona, M. G., & Medintz, I. L. (2014). Understanding enzymatic acceleration at nanoparticle interfaces: Approaches and challenges. *Nano Today, 9*, 102–131.

Kim, J., Grate, J. W., & Wang, P. (2006). Nanostructures for enzyme stabilization. *Chemical Engineering Science, 61*, 1017–1026.

Kim, Y., Ho, S. O., Gassman, N. R., Korlann, Y., Landorf, E. V., Collart, F. R., et al. (2008). Efficient site-specific labeling of proteins via cysteines. *Bioconjugate Chemistry, 19*, 786–791.

Knudsen, B. R., Jepsen, M. L., & Ho, Y. P. (2013). Quantum dot-based nanosensors for diagnosis via enzyme activity measurement. *Expert Review of Molecular Diagnostics, 13*, 367–375.

Lakowicz, J. R. (2006). *Principles of fluorescence spectroscopy* (3rd ed.). New York: Springer.
Lata, J. P., Gao, L., Mukai, C., Cohen, R., Nelson, J. L., Anguish, L., et al. (2015). Effects of nanoparticle size on multilayer formation and kinetics of tethered enzymes. *Bioconjugate Chemistry, 26*, 1931–1938.
Medintz, I. L., Clapp, A. R., Brunel, F. M., Tiefenbrunn, T., Uyeda, H. T., Chang, E. L., et al. (2006). Proteolytic activity monitored by FRET through quantum-dot peptide conjugates. *Nature Materials, 5*, 581–589.
Medintz, I. L., Clapp, A. R., Mattoussi, H., Goldman, E. R., Fisher, B., & Mauro, J. M. (2003). Self-assembled nanoscale biosensors based on quantum dot FRET donors. *Nature Materials, 2*, 630–638.
Medintz, I., & Hildebrandt, N. (2014). *FRET—Förster resonance energy transfer: From theory to applications*. Weinheim, Germany: Wiley-VCH Verlag GmbH.
Medintz, I. L., Konnert, J. H., Clapp, A. R., Stanish, I., Twigg, M. E., Mattoussi, H., et al. (2004). A fluorescence resonance energy transfer derived structure of a quantum dot-protein bioconjugate nanoassembly. *Proceedings of the National Academy of Sciences of the United States of America, 101*, 9612–9617.
Medintz, I. L., Uyeda, H. T., Goldman, E. R., & Mattoussi, H. (2005). Quantum dot bioconjugates for imaging, labelling and sensing. *Nature Materials, 4*, 435–446.
Mudhivarthi, V. K., Cole, K. S., Novak, M. J., Kipphut, W., Deshapriya, I. K., Zhou, Y., et al. (2012). Ultra-stable hemoglobin-poly(acrylic) acid conjugates. *Journal of Materials Chemistry, 22*, 20423–20433.
Mudhuvarthi, V. K., Bhambhani, A., & Kumar, C. V. (2007). Novel enzyme/DNA/inorganic nanomaterials: A new generation of biocatalysts. *Dalton Transactions, 47*, 5483–5497.
Nagy, A., Gemmill, K. B., Delehanty, J. B., Medintz, I. L., & Sapsford, K. E. (2014). Peptide-functionalized quantum dot biosensors. *IEEE Journal of Selected Topics in Quantum Electronics, 20*, 6900512.
Oh, E., Fatemi, F. K., Currie, M., Delehanty, J. B., Pons, T., Fragola, A., et al. (2013). PEGylated luminescent gold nanoclusters: Synthesis, characterization, bioconjugation, and application to one- and two-photon cellular imaging. *Particle and Particle Systems Characterization, 30*, 453–466.
Petryayeva, E., & Algar, W. R. (2013). Proteolytic assays on quantum-dot-modified paper substrates using simple optical readout platforms. *Analytical Chemistry, 85*, 8817–8825.
Petryayeva, E., Algar, W. R., & Medintz, I. L. (2013). Quantum dots in bioanalysis: A review of applications across various platforms for fluorescence spectroscopy and imaging. *Applied Spectroscopy, 67*, 215–252.
Pfeiffer, C., Rehbock, C., Hühn, D., Carrillo-Carrion, C., De Aberasturi, D. J., Merk, V., et al. (2014). Interaction of colloidal nanoparticles with their local environment: The (ionic) nanoenvironment around nanoparticles is different from bulk and determines the physico-chemical properties of the nanoparticles. *Journal of the Royal Society Interface. 11.* http://dxdoi.org/10.1098/rsif.2013.0931.
Prasuhn, D. E., Blanco-Canosa, J. B., Vora, G. J., Delehanty, J. B., Susumu, K., Mei, B. C., et al. (2010). Combining chemoselective ligation with polyhistidine-driven self-assembly for the modular display of biomolecules on quantum dots. *ACS Nano, 4*, 267–278.
Prasuhn, D. E., Deschamps, J. R., Susumu, K., Stewart, M. H., Boeneman, K., Blanco-Canosa, J. B., et al. (2010). Polyvalent display and packing of peptides and proteins on semiconductor quantum dots: Predicted versus experimental results. *Small, 6*, 555–564.
Rameshwar, T., Samal, S., Lee, S., Kim, S., Cho, J., & Kim, I. S. (2006). Determination of the size of water-soluble nanoparticles and quantum dots by field-flow fractionation. *Journal of Nanoscience and Nanotechnology, 6*, 2461–2467.
Sapsford, K. E., Algar, W. R., Berti, L., Gemmill, K. B., Casey, B. J., Oh, E., et al. (2013). Functionalizing nanoparticles with biological molecules: Developing chemistries that facilitate nanotechnology. *Chemical Reviews, 113*, 1904–2074.

Sapsford, K. E., Farrell, D., Steven Sun, S., Rasooly, A., Mattoussi, H., & Medintz, I. L. (2009). Monitoring of enzymatic proteolysis on a electroluminescent-CCD microchip platform using quantum dot-peptide substrates. *Sensors & Actuators B: Chemical, 139*, 13–21.

Shahidi, F., & Janak Kamil, Y. V. A. (2001). Enzymes from fish and aquatic invertebrates and their application in the food industry. *Trends in Food Science & Technology, 12*, 435–464.

Susumu, K., Oh, E., Delehanty, J. B., Blanco-Canosa, J. B., Johnson, B. J., Jain, V., et al. (2011). Multifunctional compact zwitterionic ligands for preparing robust biocompatible semiconductor quantum dots and gold nanoparticles. *Journal of the American Chemical Society, 133*, 9480–9496.

Susumu, K., Oh, E., Delehanty, J. B., Pinaud, F., Gemmill, K. B., Walper, S., et al. (2014). A new family of pyridine-appended multidentate polymers as hydrophilic surface ligands for preparing stable biocompatible quantum dots. *Chemistry of Materials, 26*, 5327–5344.

Tsai, P.-C., Fox, N., Bigley, A. N., Harvey, S. P., Barondeau, D. P., & Raushel, F. M. (2012). Enzymes for the homeland defense: Optimizing phosphotriesterase for the hydrolysis of organophosphate nerve agents. *Biochemistry, 51*, 6463–6475.

Walper, S. A., Turner, K. B., & Medintz, I. L. (2015). Enzymatic bioconjugation of nanoparticles: Developing specificity and control. *Current Opinion in Biotechnology, 34*, 232–241.

Wu, M., & Algar, W. R. (2015). Acceleration of proteolytic activity associated with selection of thiol ligand coatings on quantum dots. *ACS Applied Materials & Interfaces, 7*, 2535–2545.

Wu, M., Petryayeva, E., Medintz, I. L., & Algar, W. R. (2014). Quantitative measurement of proteolytic rates with quantum dot-peptide substrate conjugates and Förster resonance energy transfer. *Methods in Molecular Biology, 1199*, 215–239.

Zaks, A. (2001). Industrial biocatalysis. *Current Opinion in Chemical Biology, 5*, 130–136.

Zhu, X., Hu, J., Zhao, Z., Sun, M., Chi, X., Wang, X., et al. (2015). Kinetic and sensitive analysis of tyrosinase activity using electron transfer complexes: In vitro and intracellular study. *Small, 11*, 862–870.

CHAPTER THREE

Intense PEGylation of Enzyme Surfaces: Relevant Stabilizing Effects

**S. Moreno-Pérez, A.H. Orrego, M. Romero-Fernández,
L. Trobo-Maseda, S. Martins-DeOliveira, R. Munilla,
G. Fernández-Lorente[1], J.M. Guisan[1]**

Institute of Catalysis, Spanish Research Council, CSIC, Madrid, Spain
[1]Corresponding authors: e-mail address: gflorente@icp.csic.es; jmguisan@icp.csic.es

Contents

1. Introduction	56
2. Theory	57
3. Protocols	62
3.1 PEGylation of Chemically Aminated Enzymes	62
3.2 PEGylation of Enzymes Coated with Polymers	65
4. Inactivation of Modified Enzyme Derivatives	68
5. Conclusions	70
Acknowledgments	71
References	71

Abstract

This chapter describes the physicochemical coating of the surface of immobilized enzymes with a dense layer of polyethylene glycol (PEG) to improve enzyme stability. One hypothesis is that a dense, viscous, polar PEG layer around the enzyme would enhance enzyme thermal stability, while still providing access to the active site. PEG groups were attached by using aldehyde–dextran polymers, the dextran polymers are in turn attached to the enzyme surface that have been enriched with excess primary amino groups. The enzymes themselves were initially attached onto porous solids such that they may be separated easily from the reaction mixtures for easy downstream processing and that they may be recycled to reduce the cost of the biocatalyst. The hierarchical modification of enzyme surface with three different sublayers, under chemical design, provided a rational control at several structural levels. Few methods for increasing the number of amino groups on the surface of the enzyme are described: (a) chemical amination of carboxyl residues and (b) coating of the enzyme surface with cationic polymers containing a high percentage of primary amines. Reliable protocols for the PEGylation of four different enzymes are described here. For example, lipases from *Thermomyces lanuginosa*, *Candida antarctica* B, and *Rhizomucor miehei* attached to octyl sepharose and chemically modified via PEGylation are stabilized from 7- to 50-fold when compared to the stability of the corresponding unmodified enzyme. A derivative of endoxylanase from *Trichoderma reesei*, immobilized by multipoint covalent attachment on glyoxyl agarose, is stabilized by 50-fold. Very likely, the PEG layer generated a dense, high viscosity medium surrounding the enzyme surface and this increase in viscosity around the enzyme microenvironment resists distortion of enzyme structure by heat or other denaturing agents.

1. INTRODUCTION

Enzymes exhibit excellent functional properties to be used as therapeutic agents, industrial biocatalysts, biosensors, etc. One of the serious limitations for the widespread use of enzymes in the above applications is their poor stability against moderately high temperatures, organic solvents, extreme pHs, high ionic strengths, or digestion by macrophages. Therefore, several practical methods were devised to stabilize enzymes such as site-directed mutagenesis to produce recombinant enzymes with improved stability, and the adoption of enzymes from extremophiles such as thermophilic enzymes which are stable even around the boiling point of water. However, each of these approaches has certain limitations, which are being actively addressed currently.

For example, thermophilic enzymes are functional at elevated temperatures but they are sluggish at room temperature and below. The recombinant enzymes have made only limited progress in stabilizing enzymes. Therefore, rational chemical methodologies are being sought to control enzyme stability such that these biological catalysts can be manipulated at the molecular level to suit various key applications in chemistry, therapy, diagnostics, personalized medicine, as well as industry. The chemical space accessible via these rational methods far outweighs other approaches, and chemical manipulation of enzymes based on our current understanding of specific factors that contribute to enzyme stability provides a solid chemical basis to design rational methods for enzyme stability enhancements.

Enzymes bound to specific surfaces of particles and layered nanosheets exhibited very interesting stabilizing effects: multipoint covalent immobilization on highly activated glyoxyl supports (Guisan, 1988), interfacial adsorption of lipases on hydrophobic supports (Bastida, Sabuquillo, Armisen, Guisan, et al., 1998), enzyme intercalation in layered inorganic phosphates and phosphonates (Deshapriya & Kumar, 2013; Kumar & Chaudhari, 2000; Mudhivarthi, Bhambhani, & Kumar, Kumar, 2007) or enzyme stabilization by the popular layer-by-layer method (Decher, 2012), and hierarchical self-assembly in nanomaterials (Talbert et al., 2014).

In such stabilized enzyme derivatives, a small portion of the protein surface remains in close contact with the support surface but most of the remaining protein surface is not protected and fully exposed to the solvent, which is unprotected and prone to denaturation by external agents. Obviously, in soluble enzymes the whole enzyme surface is exposed to the medium, unprotected and prone to denaturation. The development of new strategies for enzyme stabilization via massive modification of the enzyme surface shielding it from the external medium while allowing egress and ingress of substrates could promote very interesting additional stabilizing effects. When combined with enzymes positioned on porous solid surfaces or nanosheets, this could provide additional stability for the bound enzyme while retaining several advantages of solid-bound enzymes.

Most of the soluble enzymes are highly stabilized in the presence of high concentrations of polyethylene glycol (PEG; Fernández-Lorente et al., 2015). Very likely, the presence of a medium with a high viscosity may prevent the undesired changes in enzyme structure promoted by denaturing agents (e.g., high temperatures, strong basic conditions, extreme pH values) and, therefore, the stability of the soluble enzymes greatly increases. Building on this observation, the current approach is to build such a layer around enzyme to stabilize it.

2. THEORY

Bearing in mind these interesting antecedents, we propose the fixation of a very dense layer of PEG to the enzyme surface as a possible stabilization strategy (Guisan, Moreno-Pérez, Herrera, & Romero, et al., 2015) where the thickness and surface number density of the PEG layer would play a major role.

a. Massive PEGylation of the enzyme surface should generate a highly viscous layer consisting of the PEG segments surrounding the enzyme surface. This viscous layer should prevent or reduce the distortions of the enzyme surface in the presence of denaturing agents. However, the chemistry used for the attachment has to be benign and not interfere with the functioning of the biocatalyst. This layer should also not interfere with the substrate diffusion to the enzyme active site as well.

b. An additional coating of the surface of the enzyme with a very thin hyperhydrophilic layer inserted between the enzyme surface and the PEG layer may also be useful to prevent the movement of some internal hydrophobic pockets of the enzyme. These undesirable movements of hydrophobics may be promoted by heat and by the presence of a slightly nonpolar layer of PEG. Therefore, this base layer preventing the exposure of the hydrophobics could be more important in PEGylated enzymes.

c. Dextran–aldehyde as a very suitable scaffold to fix dense layers of PEG on the enzyme surface while preventing the exposure of the hydrophobic groups from the interior of the enzyme.

For the construction of these enzyme modification layers, we use aldehyde–dextran polymers (Fig. 1). Dextran–aldehyde polymers are obtained by periodic acid oxidation of dextran, which is essentially poly(glucose) can be oxidized sequentially with periodic acid, and each glucose unit yields two adjacent aldehyde groups that are very reactive and stable (Betancor, Fuentes, Delamora-Ortiz, Guisan, et al., 2005).

These aldehyde-featuring polymers can be easily attached to a highly aminated surface of the enzyme, so that the whole enzyme surface is fully covered with the hydrophilic polymer. Literature shows that such wrapping of the enzymes with highly hydrophilic polymers such as poly(acrylic) acid enhances enzyme stability by preventing the distortions of the enzyme, possibly by preventing the exposure of the hydrophobic groups from the enzyme interior, as well as by reducing the conformational entropy of the polypeptide chain (Riccardi et al., 2016). After this first coating, a number of aldehyde groups remain in the aldehyde–dextran and these aldehyde functions are reacted with an excess of PEG-NH_2 generating a second layer consisting of PEG, thus providing a two-layer protection around the enzyme (Fig. 2).

The reaction of PEG-NH_2 with aldehydes can be further improved by using dithiothreitol (DTT 10 mM) in the reaction environment (Bolivar, Lopez-Gallego, Godoy, Guisan, et al., 2009) to stabilize the Schiff's bases formed between amino and aldehyde groups or by using NaCNBH$_3$ or 2-picoline borane to completely reduce the Schiff's bases to secondary amines (Ruhaak, Steenvoorden, Koeleman, Deelder, & Wuhrer, 2010). These chemical modifications might affect the active center of the enzyme and should be blocked with large amounts of substrate or inhibitor. In the case of lipases adsorbed on hydrophobic supports, the active center is already protected by the support surface and the performance of intense PEGylation may be done very easily without loss of activity.

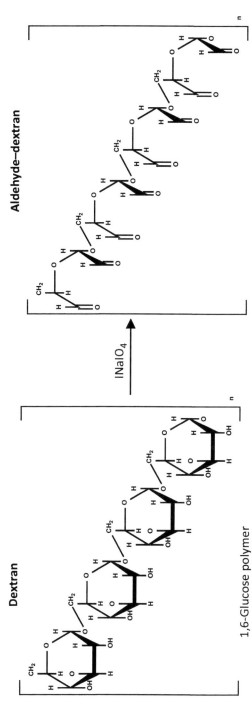

Figure 1 Preparation of aldehyde–dextran. It is obtained by periodate oxidation of dextran. Dextran is a polymer of 1,6-glucose and it is converted into a very stable polyaldehyde structure with high reactivity.

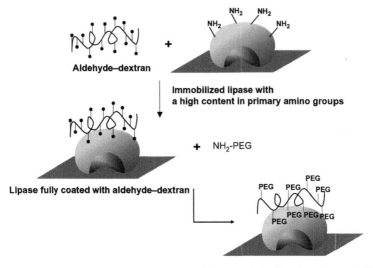

Figure 2 Aldehyde–dextran as scaffold to create a dense layer of PEG surrounding the enzyme surface. Aldehyde–dextran is attached to enzyme surfaces highly enriched in primary amino groups. The remaining aldehyde groups are modified with polyethylene glycol amine.

d. Enrichment of enzyme surface in amino groups in order to provide a complete covering of the surface with dextran–aldehyde.
   1. The distribution of Lys on enzyme surfaces is very different among proteins. The amount of these free surface Lys is even much lower when the enzyme has been previously immobilized on the support by multipoint covalent attachment, a technique that involves the region with the highest density in Lys groups (Guisan, 1988). Due to the reversibility of the amino-aldehyde bonds, the attachment of dextran–aldehyde on the enzyme surface has to occur via multipoint covalent attachment through all regions of the enzyme surface. For this reason, only very few native enzymes (those having a very high number of surface Lys residues) can be completely coated with dextran–aldehyde. Therefore, in most of the cases, enzymes should be enriched in reactive amino groups in order to improve the full coating with the dextran–aldehyde.
   2. Chemical amination of the enzyme surface. The number of surface carboxyl groups in enzymes (Asp + Glu residues) is usually much higher than the number of Lys residues. Thus, enzymes can be highly

enriched in amino groups by reaction of carboxyl groups (activated with carbodiimide) with high concentrations of ethylenediamine. In addition to that, the new amino groups are much more reactive than Lys residues. The highly aminated enzymes are easily coated with dextran–aldehyde and can even be directly PEGylated with commercial PEG modified with esters of N-hydroxysuccinimide (PEG-NHS) (Nektar Advanced PEGylation, 2005).

3. Physical amination of the enzyme surface. One elegant way to enrich enzyme surfaces in primary amino groups is the ionic adsorption of polyethyleneimine (PEI) or polyallylamine (PAA) on the ionized carboxyl groups of the enzyme (Asp + Glu residues). The multipoint ionic adsorption of these polymers on the high amount of carboxyl groups is very intense and becomes very stable (Guisan, Sabuquillo, Fernandez-Lorente, Mateo, et al., 2001). In recent study, extensive amination of glucose oxidase was shown to stabilize it more than 50-fold, even at 40 °C (Novak et al., 2015). The resulting enzyme surfaces are highly enriched in new primary amino groups, because the chemical modification or the physical adsorption with PEI introduces large numbers of primary amino groups and PAA is completely composed by primary amino groups. The physical/chemical modification of enzymes also promotes the formation of a thin hyperhydrophilic coating of the enzyme surface, preventing the direct interaction between this surface and the dense layer of PEG (Fig. 3).

The stabilizing effect of these PEG layer strongly depends on a number of variables such as the size of aldehyde–dextran, mole ratio of

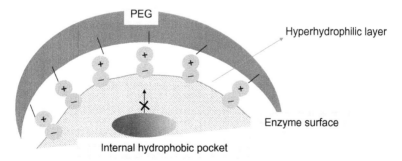

Figure 3 An ideal alternative for massive PEGylation of enzyme surface: a very thin hyperhydrophilic layer is inserted between the enzyme surface and the PEG layer. Now, the hyperhydrophilic layer may prevent the access of internal hydrophobic pockets to the enzyme surface promoted by heat or by the presence of a fairly nonpolar layer of PEG.

aldehyde–dextran to the enzyme, the extent of modification of aldehyde–dextran with amino-PEG, and the molecular weight of PEI and PAA. Here, we report key data on enzyme PEGylations with significant improvements for each enzyme and each denaturing agent.

e. Protection of the enzyme active center during physicochemical modification.

It is necessary to prevent the modification of the active center of the enzyme during these various reactions. The active center is protected in the presence of substrates or inhibitors. For example, xylan is used to protect the active center of endoxylanases. The active center of lipases adsorbed on hydrophobic supports is already protected by the support surface where the open active center is adsorbed facing the support. From this point of view, lipases are very good models to test this new stabilization strategy.

3. PROTOCOLS

Equipment
- Basic Sartorius, balance.
- Hamilton Slimtrade, pH electrode.
- Grant JV Nova, thermal water bath.
- Stuart roller mixer.
- Jasco V-630, spectrophotometer provided with magnetic stirring of the cuvettes in order to assay immobilized derivatives.

3.1 PEGylation of Chemically Aminated Enzymes

3.1.1 Rhizomucor miehei *Lipase*

Reagents
- Octyl sepharose.
- 5 mM Phosphate buffer, pH 7.
- 25 mM Phosphate buffer, pH 7.
- 0.4 mM 4-Nitrophenyl butyrate ($C_{10}H_{11}NO_4$) (*p*NPB).
- 1 M Ethylenediamine ($NH_2CH_2CH_2NH_2$), pH 4.75 (EDA).
- 5 mM N-(3-Dimethylaminopropyl)-N′-ethylcarbodiimide hydrochloride ($C_8H_{17}N_3 \cdot HCl$) (EDAC).

- 2.78 mM Dextran $((C_6H_{10}O_5)_n)$ 25,000 Da in 100 mM phosphate buffer, pH 7.
- 0.4 M Sodium periodate $(NaIO_4)$ in deionized water.
- 8.3 M Methoxypolyethylene glycol amine $(H_2NCH_2CH_2(OCH_2CH_2)_n OCH_3)$ 2000 Da (PEG).
- 100 mM Bicarbonate buffer, pH 8.5.
- 1 mg/mL Sodium borohydride $(NaCNBH_3)$ in bicarbonate buffer pH 9.
- Dialysis tubing MWCO 12,000–14,000 Da.
- Deionized water.

Method

The lipase was immobilized on octyl sepharose by interfacial activation at a low ionic strength (5 mM) in sodium phosphate buffer at 25 °C and pH 7. Periodically, the activity of the suspension and supernatant was measured using the pNPB assay (Bastida et al., 1998). The enzyme loading of the immobilized preparations was 2 mg of lipase per gram of support. One gram of immobilized lipase derivative was added to a 10 mL solution containing 1 M EDA, pH 4.75. Solid EDAC was added to a final concentration of 5 mM. After 90 min of gentle agitation at room temperature, the aminated immobilized lipase derivative was vacuum filtered and thoroughly washed with 50 mL of distilled water five times.

Dextran (10.02 g) of molecular weight of 25,000 Da was diluted in 300 mL of distilled water. In order to modify this polymer to 100% of aldehyde groups, it was completely oxidized by the addition of 26.16 g of solid sodium periodate. After 90 min of gentle stirring at room temperature, the mixture was dialyzed against 2 L of distilled water. Then, for pH adjustment, 300 mL dialyzed solution was mixed with 300 mL of 100 mM sodium phosphate buffer, pH 7. Immobilized lipase derivative (1 g) was added to 10 mL of dextran–aldehyde solution. After 16 h of gentle stirring at 4 °C, the immobilized lipase-dextran derivative was vacuum filtered. One gram of immobilized lipase–dextran derivative was added to 15 mL solution of 8.3 M of methoxypolyethylene glycol amine 2000 Da and 100 mM bicarbonate buffer 100 mM, pH 8.5. After 2 h of gentle stirring at room temperature immobilized lipase–dextran–PEG was reduced at pH 8.5 by adding 15 mg of solid sodium borohydride for 30 min at 25 °C. Finally, the modified derivative was filtered and thoroughly washed with 50 mL of distilled water five times and stored at 4 °C. The final modified derivative retained 100% of initial activity compared with soluble lipase provided.

3.1.2 Thermomyces lanuginosa *Lipase*
Reagents
- Octyl sepharose from GE Healthcare.
- 5 mM Phosphate buffer, pH 7.
- 25 mM Phosphate buffer, pH 7.
- 0.4 mM 4-Nitrophenyl butyrate ($C_{10}H_{11}NO_4$) (*p*NPB).
- 1 M Ethylenediamine ($NH_2CH_2CH_2NH_2$), pH 4.75 (EDA).
- 10 mM *N*-(3-Dimethylaminopropyl)-*N'*-ethylcarbodiimide hydrochloride ($C_8H_{17}N_3 \cdot HCl$) (EDAC).
- 0.41 mM Dextran ($[C_6H_{10}O_5]_n$) 40,000 Da in 100 mM phosphate buffer, pH 7.
- 0.081 M Sodium periodate ($NaIO_4$) in deionized water.
- 20 mM Sodium periodate ($NaIO_4$) in deionized water.
- 8.3 M Methoxypolyethylene glycol amine ($H_2NCH_2CH_2(OCH_2CH_2)_n OCH_3$) 2000 Da (PEG).
- 100 mM Bicarbonate buffer, pH 8.5.
- 1 mg/mL Sodium borohydride ($NaCNBH_3$) in bicarbonate buffer, pH 9.
- Dialysis tubing MWCO 12,000–14,000 Da.
- Deionized water.

Method

The lipase was immobilized on octyl sepharose by interfacial activation at a low ionic strength (5 mM) in sodium phosphate buffer at 25 °C and pH 7 (Bastida et al., 1998). Periodically, the activity of the suspension and supernatant was measured using the *p*NPB assay. The enzyme load of the immobilized preparations was 2 mg of lipase per gram of support.

A total of 1 g of immobilized lipase derivative was added to a 10 mL solution containing 1 M EDA, pH 4.75. Solid EDAC was added to a final concentration of 10 mM. After 90 min of gentle agitation at room temperature, the aminated immobilized lipase derivative was filtered and thoroughly washed with 50 mL of distilled water, five times.

A total of 10.02 g of dextran with a molecular weight of 40,000 Da were diluted in 300 mL of distilled water. In order to modify this polymer to have upto 20% of aldehyde groups, it was oxidized by the addition of 5.232 g of solid sodium periodate. After 90 min of gentle stirring at room temperature, the mixture was dialyzed against 2 L of distilled water.

Then, for pH adjustment, the 300 mL dialyzed solution was mixed with 300 mL of 100 mM sodium phosphate buffer, pH 7. One gram of immobilized lipase derivative was added to 10 mL of dextran–aldehyde solution. After 16 h of gentle stirring at 4 °C, the immobilized lipase derivative was reduced with 1 mg/mL sodium periodate for 30 min at room temperature, and then vacuum filtered and washed with distilled water. One gram of immobilized lipase–dextran derivative was added to 10 mL of 20 M sodium periodate solution for 2 h at 25 °C to produce new aldehyde groups, and then vacuum filtered and washed with distilled water. One gram of immobilized lipase–dextran derivative was added to 15 mL solution of 8.3 M methoxypolyethylene glycol amine 2000 Da and 100 mM bicarbonate buffer, pH 8.5. After 2 h of gentle stirring at room temperature, immobilized lipase–dextran–PEG was reduced at pH 8.5 by adding 15 mg of solid sodium borohydride over 30 min at 25 °C. Finally, they were washed several times with 100 mL of distilled water and stored at 4 °C. The final modified derivative retained 100% of initial activity when compared with the activity of soluble lipase.

3.2 PEGylation of Enzymes Coated with Polymers
3.2.1 Candida antarctica *B Lipase*
Reagents
- Octyl sepharose.
- 5 mM Phosphate buffer, pH 7.
- 25 mM Phosphate buffer, pH 7.
- 25 mM Phosphate buffer, pH 8.
- 25 mM Bicarbonate buffer, pH 8.5.
- 0.4 mM 4-Nitrophenyl butyrate ($C_{10}H_{11}NO_4$) (*p*NPB).
- 4 mM Polyethylenimine 25,000 Da (($CH_2CH_2NH)_n$) (PEI).
- 1.34 mM Dextran (($C_6H_{10}O_5)_n$) 25,000 Da in deionized water.
- 0.33 M Sodium periodate ($NaIO_4$) in deionized water.
- 20 mM Sodium periodate ($NaIO_4$) in deionized water.
- 8.3 M Methoxypolyethylene glycol amine ($H_2NCH_2CH_2(OCH_2CH_2)_n OCH_3$) 2000 Da (PEG).
- 100 mM Bicarbonate buffer, pH 8.5.
- 1 mg/mL Sodium borohydride ($NaCNBH_3$) in bicarbonate buffer, pH 9.
- Dialysis tubing MWCO 12,000–14,000 Da.
- Deionized water.

Method

The lipase was immobilized on octyl sepharose by interfacial activation at a low ionic strength (5 mM) in sodium phosphate buffer at 25 °C and pH 7 (Bastida et al., 1998). Periodically, the activity of the suspension and supernatant was measured using the pNPB assay. The enzyme load of the immobilized preparations was 2 mg of lipase per gram of support.

A total of 2 g of polyethylenimine 25,000 Da was added to 20 mL of 25 mM sodium phosphate buffer pH 8. After being adjusted to pH 8, this polyethylenimine solution was added to 1 g of immobilized lipase derivative and incubated at 4 °C overnight under gentle agitation. Finally, it was vacuum filtered and the resulting modified immobilized lipase derivative thoroughly washed with 50 mL of distilled water five times.

A total of 10.02 g of dextran with a molecular weight of 25,000 Da were diluted in 300 mL of distilled water. In order to modify this polymer up to 80% of aldehyde groups, it was oxidized by the addition of 20.928 g of solid sodium periodate. After 90 min of gentle stirring at room temperature, the mixture was dialyzed against 2 L of distilled water. Then, for pH adjustment, the 300 mL dialyzed solution was mixed with 300 mL of 100 mM sodium phosphate buffer, pH 7. Immobilized lipase derivative (1 g) was added to 10 mL of dextran–aldehyde solution. After 16 h of gentle stirring at 4 °C, the immobilized lipase derivatives were vacuum filtered. One gram of immobilized lipase–dextran derivative was added to 15 mL solution of methoxypolyethylene glycol amine 2000 Da 4.15 M and bicarbonate buffer 100 mM, pH 8.5. After overnight stirring at room temperature, immobilized lipase–dextran–PEG was reduced at pH 8.5 with 15 mg of solid sodium borohydride for 30 min at 25 °C. Finally, they were washed several times with around 100 mL of distilled water and stored at 4 °C. The final modified derivative retained 100% of initial activity when compared to the activity of soluble lipase.

3.2.2 Bioxilanase L Plus (Trichoderma reesei endo-1,4-β-Xylanase)
Reagents
- Glyoxyl agarose 10 BCL.
- 100 mM Bicarbonate buffer, pH 10.
- 25 mM Bicarbonate buffer, pH 8.5.
- 25 mM Phosphate buffer, pH 8.
- 25 mM Phosphate buffer, pH 7.
- 43.8 M 3,5-Dinitrosalicylic acid (($O_2N)_2C_6H_2$-2-(OH)CO_2H) (DNS).

- 1% (w/v) Xylan from beechwood in 50 mM sodium acetate buffer at pH 5.0.
- 45 nM Polyethylenimine 10,000 Da (($CH_2CH_2NH)_n$).
- 5.57 mM Dextran (($C_6H_{10}O_5)_n$) 6000 Da in deionized water.
- 0.4 M Sodium periodate ($NaIO_4$) in deionized water.
- 8.3 M Methoxypolyethylene glycol amine ($H_2NCH_2CH_2(OCH_2CH_2)_n$ OCH_3) 2000 Da (PEG).
- 100 mM Bicarbonate buffer, pH 8.5.
- 1 mg/mL Sodium borohydride ($NaCNBH_3$) in bicarbonate buffer, pH 9.
- Mixture of 1% (w/v) xylan from beechwood in deionized water.
- Dialysis tubing MWCO 12,000–14,000 Da.
- Deionized water.

Method

The xylanase was immobilized for 4 h on 10 BCL aldehyde–agarose gel by multicovalent attachment in 100 mM bicarbonate buffer at 25 °C and pH 10 (Guisan, 1988). Periodically, the activity of the suspension and supernatant was measured using the DNS assay (Miller, 1959). The enzyme load of the immobilized preparations was 9 mg of xylanase per gram of support. A total of 9 mg of 10,000 Da polyethylenimine was added to 20 mL of 25 mM sodium phosphate buffer, pH 8. After being adjusted to pH 8, this polyethylenimine solution was added to 1 g of immobilized xylanase derivative and incubated at 4 °C overnight under gentle agitation. Finally, it was vacuum filtered and the resulting modified immobilized xylanase derivative washed with 50 mL of distilled water five times.

A total of 10.02 g of dextran with a molecular weight of 10,000 Da were diluted in 300 mL of distilled water. In order to modify this polymer up to 100% of aldehyde groups, it was completely oxidized by the addition of 26.16 g of sodium periodate. After 90 min of gentle stirring at room temperature, the mixture was dialyzed against 2 L of distilled water. Then, for pH adjustment, the 300 mL dialyzed solution was mixed with 300 mL of 100 mM sodium phosphate buffer pH 7. Immobilized xylanase derivative (1 g) was added to 10 mL of dextran–aldehyde solution. After 16 h of gentle stirring at 4 °C, the immobilized xylanase–dextran derivative was vacuum filtered. One gram of immobilized xylanase–dextran derivative was added to 15 mL solution of methoxypolyethylene glycol amine 2000 Da 3.32 M and bicarbonate buffer 100 mM, pH 8.5. After overnight of gentle stirring at 4 °C, immobilized xylanase–dextran–PEG was reduced in the presence of 1% (w/v) of substrate (xylan) at pH 8.5 with 15 mg of solid sodium borohydride for 30 min at 25 °C. Finally, they were washed several times with around 100 mL of distilled water and stored at 4 °C. The final modified derivative retained 100% of initial activity when compared to the activity of the soluble xylanase.

4. INACTIVATION OF MODIFIED ENZYME DERIVATIVES

The thermal inactivation of PEGylated derivatives in comparison with the unmodified ones was tested. The inactivation was carried out by suspension of the 1 g of enzyme derivatives (20 mg lipase and 9 mg xylanase per gram of support) in 10 mL of 25 mM phosphate buffer pH 7.0 and the suspension was incubated at 50, 60, or 70 °C. At different times, samples of the suspensions were withdrawn and their activities were determined: lipases were assayed with pNPB as substrate, as previously described (Fernandez-Lorente, Godoy, Lopez-Gallego, Guisan, et al., 2008). Endoxylanase was assayed by hydrolysis of xylan to release reducing sugars as previously described (Aragón, Mateo, Ruiz-Matute, Guisan, et al., 2013). The remaining activities were calculated as the ratio between activity at a given time and activity at zero time of incubation (Figs. 4–7).

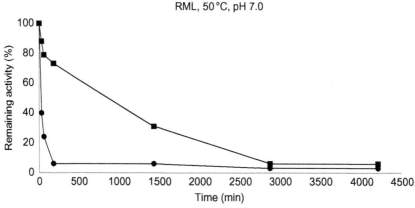

Figure 4 Time course of thermal inactivation of two RML derivatives at 50 °C and pH 7.0. The thermal inactivation of PEGylated derivatives was compared with the unmodified ones. The inactivation was carried out using 1 g of enzyme derivatives (20 mg lipase) in 10 mL of 25 mM phosphate buffer, pH 7.0, and incubated at 50 °C. At different times, samples of the suspensions were withdrawn and activities determined: lipase was assayed with pNPB as substrate, (Fernandez-Lorente et al., 2008). The retention activities were calculated as the ratio of activity at a given time and activity at zero time of incubation (expressed as percentages). Circles: unmodified lipase (RML) derivatives. Squares: lipase (RML) derivative coated with PEG.

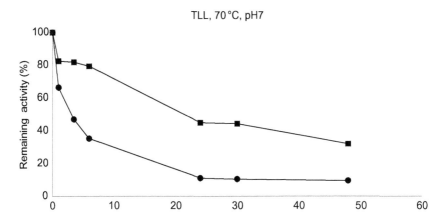

Figure 5 Time course of thermal inactivation of two *Thermomyces lanuginosa* lipase (TLL) derivatives at 70 °C and pH 7.0. The thermal inactivation of PEGylated lipase derivatives in comparison with the unmodified ones. The inactivation was carried out using 1 g of lipase (TLL) derivatives (20 mg lipase per gram of support) in 10 mL of 25 mM phosphate buffer, pH 7.0, and the suspension incubated at 70 °C. At different times, samples were withdrawn and their remaining activities determined: lipases were assayed with pNPB as substrate (Fernandez-Lorente et al., 2008). The retention of activities were calculated as the ratio of activity at a given time and activity at zero time of incubation (expressed as percentages). Circles: unmodified lipase derivative. Squares: lipase derivative coated with PEG.

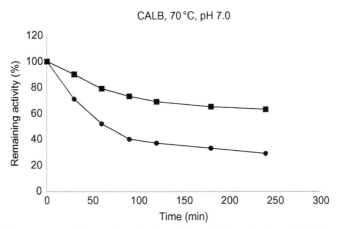

Figure 6 Time course of thermal inactivation of two CALB derivatives at 70 °C and pH 7.0. The thermal inactivation of PEGylated derivatives in comparison with the unmodified ones. The inactivation was carried out using suspension of the 1 g of enzyme derivatives (20 mg lipase (CALB) per gram of support) in 10 mL of 25 mM phosphate buffer pH 7.0 and incubated at 70 °C. At different times, samples of the suspensions were withdrawn and their activities determined: lipases were assayed with pNPB as substrate (Fernandez-Lorente et al., 2008). The retained activities were calculated as the ratio between activity at a given time and activity at zero time of incubation (expressed as percentages). Circles: unmodified lipase derivative. Squares: lipase derivative coated with PEG.

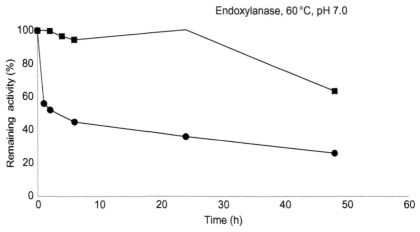

Figure 7 Time course of thermal inactivation of two endoxylanase derivatives at 60 °C and pH 7.0. The thermal inactivation of PEGylated derivatives in comparison to the unmodified ones was tested. The inactivation was carried out by suspension of the 1 g of enzyme derivatives (9 mg xylanase per gram of support) in 10 mL of 25 mM phosphate buffer pH 7.0 and the suspension was incubated at 60 °C. At different times, samples of the suspensions were withdrawn and their remaining activities determined: endoxylanase was assayed by hydrolysis of xylan to release reducing sugars (Aragón et al., 2013). The retention of activities were calculated as the ratio between activity at a given time and activity at zero time of incubation (expressed as percentages). Circles: unmodified endoxylanase derivative. Squares: endoxylanase derivative coated with PEG.

5. CONCLUSIONS

Intense PEGylation of enzyme surfaces seems to promote interesting stabilizing effects. Key examples are described here with improved thermal stabilization ranging from 7- to 50-fold, when compared to the stabilities of the corresponding unmodified enzymes. These stabilizations were dramatically higher than that achieved by controlled immobilization techniques (e.g., multipoint covalent immobilization) but not enzyme intercalation in layered solids. We also found that the thermal stabilities are much greater at lower temperatures, where the PEG layers become much more viscous. Current study gives an insight into the design of thermally stable enzymes via protein–polymer conjugates by controlling the mode and the number of polymer to protein linkages introduced.

ACKNOWLEDGMENTS
This work was sponsored by the Spanish Ministry of Science and Innovation (projects AGL-2009-07526 and BIO2012-36861) and by the European Commission (C-KBBE/3293). We gratefully recognize the Spanish Ministry of Science and Innovation for the "Ramón y Cajal" contract for Dr. Fernandez-Lorente and for the FPI contract to Sonia Moreno-Perez. We would like to thank Novozymes and Ramiro Martinez for the generous gift of commercial lipases.

REFERENCES
Aragón, C. C., Mateo, C., Ruiz-Matute, A. I., Guisan, J. M., et al. (2013). Production of xylo-oligosaccharides by immobilized-stabilized derivatives of endo-xylanase from *Streptomyces halstedii*. *Process Biochemistry*, *48*(3), 478–483.

Bastida, A., Sabuquillo, P., Armisen, P., Guisan, J. M., et al. (1998). A single step purification, immobilization, and hyperactivation of lipases via interfacial adsorption on strongly hydrophobic supports. *Biotechnology and Bioengineering*, *58*(5), 486–493.

Betancor, L., Fuentes, M., Delamora-Ortiz, G., Guisan, J. M., et al. (2005). Dextran aldehyde coating of glucose oxidase immobilized on magnetic nanoparticles prevents its inactivation by gas bubbles. *Journal of Molecular Catalysis B: Enzymatic*, *32*(3), 97–101.

Bolivar, J. M., Lopez-Gallego, F., Godoy, C., Guisan, J. M., et al. (2009). The presence of thiolated compounds allows the immobilization of enzymes on glyoxyl agarose at mild pH values: New strategies of stabilization by multipoint covalent attachment. *Enzyme and Microbial Technology*, *45*(6–7), 477–483.

Decher, G. (2012). Layer-by-layer assembly (putting molecules to work). In G. Decher & J. B. Schlenoff (Eds.), *Multilayer thin films: Sequential assembly of nanocomposite materials* (2nd ed.). Germany: Wiley-VCH Verlag GmbH Co.

Deshapriya, I. K., & Kumar, C. V. (2013). Nano-bio interfaces: Charge control of enzyme/inorganic interfaces for advanced biocatalysis. *Langmuir*, *29*, 14001–14016.

Fernandez-Lorente, G., Godoy, C., Lopez-Gallego, F., Guisan, J. M., et al. (2008). Solid-phase chemical amination of a lipase from *Bacillus thermocatenulatus* to improve its stabilization via covalent immobilization on highly activated glyoxyl-agarose. *Biomacromolecules*, *9*(9), 2553–2561.

Fernández-Lorente, G., Lopez-Gallego, F., Bolivar, J. M., Rocha-Martin, J., Moreno-Perez, S., & Guisan, J. M. (2015). Immobilization of proteins on highly activated glyoxyl supports: Dramatic increase of the enzyme stability via multipoint immobilization on pre-existing carriers. *Current Organic Chemistry*, *19*(17), 1719–1731.

Guisan, J. M. (1988). Aldehyde-agarose gels as activated supports for immobilization-stabilization of enzymes. *Enzyme and Microbial Technology*, *10*(6), 375–382.

Guisan, J. M., Moreno-Pérez, S., Herrera, A., Romero, M., et al. (2015). Enzyme stabilization by intense pegilation of the enzyme surface. Spanish Patent P201531630.

Guisan, J. M., Sabuquillo, P., Fernandez-Lorente, G., Mateo, C., et al. (2001). Preparation of new lipases derivatives with high activity–stability in anhydrous media: Adsorption on hydrophobic supports plus hydrophilization with polyethylenimine. *Journal of Molecular Catalysis B: Enzymatic*, *11*(4–6), 817–824.

Kumar, C. V., & Chaudhari, A. (2000). Proteins immobilized at the galleries of alpha-zirconium phosphates: Structure and activity studies. *Journal of the American Chemical Society*, *122*, 830–837.

Miller, G. L. (1959). Use of dinitrosalicylic acid reagent for determination of reducing sugar. *Analytical Chemistry*, *31*, 426–428.

Mudhivarthi, V. K., Bhambhani, A., & Kumar, C. V. (2007). Novel enzyme/DNA/ inorganic nanomaterials: a new generation of biocatalysts. *Dalton Transactions*, (47), 5483–5497.

Nektar Advanced PEGylation. (2005). *Commercial information from Nektar therapeutics company*.

Novak, M. J., Pattammattel, A., Koshmerl, B., Puglia, M., Williams, C., & Kumar, C. V. (2015). 'Stable-on-the-table' enzymes: Engineering enzyme-graphene oxide interface for unprecedented stability of the biocatalysts. *ACS Catalysis*, *6*, 339–447.

Riccardi, C. M., Mistri, D., Hart, O., Anuganti, M., Lin, Y., Kasi, R. M., & Kumar, C. V. (2016). Covalent interlocking of glucose oxidase and peroxidase in the voids of paper: enzyme–polymer "spider webs". *Chemical Communications*, *52*, 2593–2596. http://dx.doi.org/10.1039/C6CC00037A.

Ruhaak, L. R., Steenvoorden, E., Koeleman, C. A. M., Deelder, A. M., & Wuhrer, M. (2010). 2-Picoline-borane: A non-toxic reducing agent for oligosaccharide labeling by reductive amination. *Proteomics*, *10*, 2330–2336.

Talbert, J. N., Wang, L.-S., Duncan, B., Jeong, Y., Andler, S. M., Rotello, V. M., et al. (2014). Immobilization and stabilization of lipase (CaLB) through hierarchical interfacial assembly. *Biomacromolecules*, *15*, 3915–3922.

CHAPTER FOUR

Immobilization of Lipases on Heterofunctional Octyl–Glyoxyl Agarose Supports: Improved Stability and Prevention of the Enzyme Desorption

N. Rueda[*,†], J.C.S. dos Santos[*,‡], R. Torres[†,1], C. Ortiz[§], O. Barbosa[¶], R. Fernandez-Lafuente[*,2]

[*]Departamento de Biocatálisis, Instituto de Catálisis-CSIC, Campus UAM-CSIC, Madrid, Spain
[†]Escuela de Química, Grupo de investigación en Bioquímica y Microbiología (GIBIM), Edificio Camilo Torres 210, Universidad Industrial de Santander, Bucaramanga, Colombia
[‡]Departamento de Engenharia Química, Universidade Federal Do Ceará, Campus Do Pici, Fortaleza, Ceará, Brazil
[§]Escuela de Bacteriología y Laboratorio Clínico, Universidad Industrial de Santander, Bucaramanga, Colombia
[¶]Departamento de Química, Facultad de Ciencias, Universidad del Tolima, Ibagué, Colombia
[2]Corresponding author: e-mail address: rfl@icp.csic.es

Contents

1. Theory 74
2. Equipment 76
3. Materials 76
   3.1 Solutions 76
4. Step 1. Preparation of the Support Octyl–Glyoxyl Agarose 77
   4.1 Overview 77
5. Step 2. Immobilization of Lipases via Interfacial Activation on Octyl–Glyoxyl Agarose 79
   5.1 Overview 79
6. Step 3. Covalent Immobilization of Adsorbed Lipases on Octyl–Glyoxyl Agarose 82
   6.1 Overview 82
Acknowledgments 84
References 84

Abstract

Lipases are among the most widely used enzymes in industry. Here, a novel method is described to rationally design the support matrix to retain the enzyme on the support matrix without leaching and also activate the enzyme for full activity retention. Lipases

---

[1] Current address: Laboratorio de Biotecnología, Instituto Colombiano del Petróleo-Ecopetrol, Piedecuesta, Bucaramanga, Colombia.

are interesting biocatalysts because they show the so-called interfacial activation, a mechanism of action that has been used to immobilize lipases on hydrophobic supports such as octyl–agarose. Thus, adsorption of lipases on hydrophobic surfaces is very useful for one step purification, immobilization, hyperactivation, and stabilization of most lipases. However, lipase molecules may be released from the support under certain conditions (high temperature, organic solvents), as there are no covalent links between the enzyme and the support matrix. A heterofunctional support has been proposed in this study to overcome this problem, such as the heterofunctional glyoxyl–octyl agarose beads. It couples the numerous advantages of the octyl–agarose support to covalent immobilization and creates the possibility of using the biocatalyst under any experimental conditions without risk of enzyme desorption and leaching. This modified support may be easily prepared from the commercially available octyl–agarose. Preparation of this useful support and enzyme immobilization on it via covalent linking is described here. The conditions are described to increase the possibility of achieving at least one covalent attachment between each enzyme molecule and the support matrix.

1. THEORY

Lipases are among the most widely used enzymes in biocatalysis, both in academia and industry. These enzymes are stable in different reaction media, and they can catalyze many different reactions (they are outstanding as promiscuous enzymes) recognizing very different substrates and at the same time exhibiting a very high regio and/or enantio selectivity and/or specificity (Jaeger & Eggert, 2002; Jaeger & Reetz, 1998; Schmid & Verger, 1998).

The catalytic mechanism of lipases is complicated (Grochulski et al., 1993; Winkler, D'Arcy, & Hunziker, 1990). These enzymes exist in at least two forms, closed and open forms. In the closed form, the active site is secluded from the reaction medium by a peptide chain called flat or lid, which is usually in the inactive state. In homogenous aqueous media, this exists in equilibrium with a lipase form where the lid is open, exposing the huge hydrophobic pocket formed by the internal side of the lid and the surroundings of the active center. This open form is unstable, and the equilibrium is displaced toward the closed form. However, in the presence of a hydrophobic surface, that may be a drop of substrate (Verger, 1997), or any other hydrophobic surface, this form becomes adsorbed via this pocket (Manoel, dos Santos, Freire, Rueda, & Fernandez-Lafuente, 2015) and produces the so-called interfacial activation of lipases.

This tendency of lipases to become adsorbed on any hydrophobic surface may have some drawbacks when using lipases in free form: for example, a lipase may become adsorbed on any hydrophobic protein present in the crude protein extract or even on the open form of other lipase molecule (Palomo et al., 2003). This may complicate the characterization of soluble form of lipases.

However, this hydrophobic affinity has been used as a way to specifically immobilize lipases. Using a hydrophobic support, at low ionic strengths, lipases in their open form become adsorbed very rapidly while other water-soluble proteins presented in the crude extract are not (Fernandez-Lafuente, Armisén, Sabuquillo, Fernández-Lorente, & Guisán, 1998; Manoel et al., 2015). Thus, this immobilization protocol permits the immobilization, purification, and hyperactivation of lipases in one step. Lipases adsorbed on hydrophobic supports via interfacial activation are shown to be very stable, usually more stable than multipoint covalent attached lipase preparations (Palomo et al., 2002). Due to the strong adsorption which may involve many lipase contacts in the multipoint adsorption, such immobilized lipase may be used in a variety of media, even in media with a moderate concentration of organic solvents. However, enzyme molecules may be released from the support under certain conditions such as high concentration of organic cosolvents (Fernandez-Lorente et al., 2010) or even after lipase inactivation. This has the advantage of the enzyme being released from the support after its inactivation, under more drastic conditions, and the support may be reused. However, this also limits the range of conditions where the immobilized enzyme may be utilized, and has the risk of some enzyme leakage during the catalytic operation.

To solve the above problem of leaching, the concept of heterofunctional support has been employed (Barbosa et al., 2013). A heterofunctional support will have at least two kinds of functions in their surface that may interact with the enzyme, but not under all conditions. Thus, we can perform the immobilization under conditions where the lipase immobilization is caused by one functional group, and then change the conditions to permit other type of enzyme–support interactions, for example, a covalent bond.

In this case, we present the octyl–glyoxyl agarose as a very simple heterofunctional support useful to solve the problems associated with the octyl–support (Bernal, Illanes, & Wilson, 2014; Rueda et al., 2015; Suescun et al., 2015). Glyoxyl groups may be obtained by simple oxidation of the diols present in the agarose support. This glyoxyl group may react with the primary amino groups of proteins giving very unstable imino bonds between the enzyme and the support. This way, multiple attachments are formed for the enzyme to be incorporated onto the support. Another advantage of the imino-attachment is the pH conditions required for the attachment. The pK_a of the side chain amino groups of external Lys is around 10.7, and hence the glyoxyl supports will react with these at alkaline pH, usually 10 or even higher (Mateo et al., 2005). Therefore, glyoxyl group is ideal to develop the idea of heterofunctional support. Using neutral pH and octyl–glyoxyl supports, the first immobilization of lipases is caused by the interfacial activation of the lipase versus the layer of octyl groups, keeping the advantages of this immobilization protocol. The subsequent increase of the pH to conditions where the enzyme may covalently react with the glyoxyl groups, may permit the establishment of some covalent bonds with the support matrix and prevent leaching (Rueda et al., 2015).

2. EQUIPMENT

We have used a solid-supported biocatalyst which is suspended in a liquid, in most of the sections of this paper, and therefore, it is necessary to use stirring systems in all support activation, immobilization, or activity determination steps to avoid external diffusion limitations. The agitation may be achieved using magnetic stirring, orbital stirring, shakers, etc., although magnetic stirring may mill the agarose particles. No other special equipment is required; except a spectrophotometer and the equipment used to determine the enzyme activity which might also require a spectrophotometer, or a pHstat, HPLC, etc. During activity determinations, stirring is required to keep the suspensions from settling down. Thus, when a spectrophotometer is used for activity studies, it needs to have a stirring unit (e.g., that from JASCO). Centrifugation of the sample may be required if the beginning enzyme solution is not clear.

3. MATERIALS

Octyl-Sepharose 4BCL (octyl–agarose)
Sodium periodate ($NaIO_4$)
Sodium borohydride ($NaBH_4$)
Lipases
Potassium iodide (KI)

3.1 Solutions

Oxidation of the support: 10 mM sodium periodate.
Solution 1 to determine the periodate concentration: 10% KI (w/v).

Solution 2 to determine the periodate concentration: saturated sodium bicarbonate aqueous solution.

Immobilization buffer: 5 mM sodium phosphate or acetate at the desired pH value.

Incubation buffer: 50 mM sodium bicarbonate equilibrated at pH 10 with 1 M NaOH.

Desorption buffers: sodium phosphate or acetate at the desired pH value containing different concentrations of the detergent indicated for each enzyme.

4. STEP 1. PREPARATION OF THE SUPPORT OCTYL–GLYOXYL AGAROSE

4.1 Overview

The diols of commercial octyl–agarose 4BCL may be used to generate the glyoxyl groups. This diols are produced by the opening of the epoxy groups during cross-linking of agarose and modification of epoxy agarose with octanol to produce octyl–agarose (Fig. 1). The amount may range between

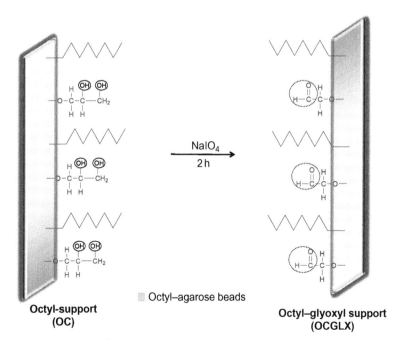

Figure 1 Preparation of octyl-glyoxyl agarose beads.

30 and 60 µmol of diols per wet gram of support. Wet support is defined as the support vacuum dried in a sintered filter eliminating the interparticle water or water loosely bound between the particles, while leaving most of the intraparticle water or the water that is bound to the support matrix.

Duration

2 h.

Protocol

1. Oxidation of octyl–agarose with sodium periodate
   1.1. Wash the octyl agarose beads five times with 5 vol of distilled water to eliminate the ethanol that the commercial sample contains.
   1.2. Put the support in a sintered glass filter connected to a flask and a vacuum pump to eliminate the interparticle water.
   1.3. Weight 105 g of octyl agarose (equivalent to 150 mL of packed volume).
   1.4. Resuspend the washed octyl agarose in 500 mL of periodate solution for 2 h under continuous mild stirring.
   1.5. After octyl agarose oxidation, keep a sample of supernatant for the determination of glyoxyl groups generated.
   1.6. Wash with an excess of distilled water.
   1.7. Keep in the refrigerator (4–8 °C).

TIP 1.1

Any reaction between a solid and dissolved molecules requires stirring. Agarose is milled using magnetic stirring, but it is highly resistant to mechanical stirring (using paddle stirrer). Moreover, orbital shakers may be also used, as it is very simple to keep the agarose in suspension.

TIP 1.2

If the support is not going to be used in a short time, storage as a suspension in 20% (v/v) ethanol to prevent drying of the matrix or to suppress microorganism contamination.

TIP 1.3

In order to avoid the risks of destroying the particles' structure due to the freezing of the intraparticle water, ensure that the temperature is never under 4 °C.

2. Determination of number of glyoxyl groups generated

    Oxidation of diols with sodium periodate is a stoichiometric reaction. Therefore, we can determine the number of glyoxyl groups formed by determining the periodate consumption during the activation step.

This protocol describes the formation of 25 μmol of glyoxyl groups per wet gram of support.

2.1. Mix 1 vol of each of the solutions 1 and 2 in a cuvette, and mark baseline correction in the spectrophotometer from 350 to 600 nm.

2.2 Add 100 μL of the sodium periodate solution, selecting the wavelength where the absorbance is 1.

2.3 Add 100 μL of the supernatant of the oxidizing solution to determine the decrease in periodate concentration.

TIP 2.1

The method is based on I_2 determination; the measures must be performed between 3 and 5 min to avoid the decrease of the signal.

5. STEP 2. IMMOBILIZATION OF LIPASES VIA INTERFACIAL ACTIVATION ON OCTYL–GLYOXYL AGAROSE

5.1 Overview

Figure 2 summarizes the immobilization steps. The first step is the interfacial activation of the lipase versus the hydrophobic layer of octyl groups (Fernandez-Lafuente et al., 1998; Manoel et al., 2015). It is very simple and rapid. The immobilization pH must be selected to ensure that the lipase is not first covalently immobilized instead of immobilized via interfacial activation. However, immobilization via interfacial activation is very rapid and that glyoxyl groups require pH 10 to immobilize enzymes (Mateo et al., 2005). Thus, it is easy to find conditions where the immobilization of lipases via covalent reactions may be irrelevant.

Duration

Depending on the lipase and level of loading desired, the required time may range from 30 min to 24 h (for full loading of the support with lipase).

Protocol

3.1. Determine the lipase activity with a clear lipase solution. As we wish that the lipase immobilization proceed via interfacial activation, check that the lipase cannot become immobilized on a glyoxyl support under the same conditions.

3.2. Add the desired amount of washed octyl–glyoxyl agarose (maximum loading depend on the enzyme, may range from 10 to 20 mg lipase/wet gram of agarose beads).

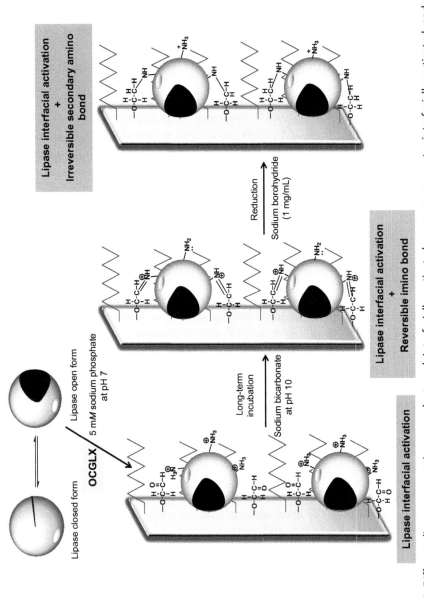

Figure 2 Different lipase preparations: covalent and interfacially activated on pure supports, interfacially activated and covalently immobilized on heterofunctional glyoxyl-octyl agarose beads.

3.3 Measure the activity of supernatant and suspension by any lipase activity determination method (spectrum, HPLC, pHstat, etc.).

3.4. When all enzyme activity is incorporated to the support or there is no significant further immobilization after 2 h, the enzyme preparation may be washed with an excess of distilled water to eliminate the non-immobilized enzyme (this protocol permits the selective adsorption of lipases) and additives.

3.5. Go to Step 3.

TIP 3.1

Ensure that the lipase solution has no precipitates to begin with to avoid mixing with the support if it is not resolubilized, closing the pores of the support, and even the pores of the filters. In case, the solution is not perfectly clear, centrifuge it to ensure elimination of any particles.

TIP 3.2

Low ionic strength is recommended (5 mM sodium phosphate at pH 7) not only to improve the selectivity of the immobilization toward lipases but also to maximize the immobilization rate.

TIP 3.3

Only multimeric enzymes or monomeric ones partially proteolyzed may become immobilized on glyoxyl supports at pH under 8. Moreover, some lipases may require alkaline pH for solubility or stability matters. In these cases, the use of buffers that reduce the glyoxyl reactivity (borate) or that can compete with the glyoxyl groups in the support (Tris, Gly, other aminated substances) may be used to prevent the covalent immobilization (Mateo et al., 2005).

TIP 3.4

Ensure stirring that avoids external diffusion limitations for activity retention, while preserving the integrity of agarose (e.g., avoid magnetic stirring).

TIP 3.5

After stirring to ensure a homogenous suspension, use cut pipette tips to permit the easy entry of the support. For the determination of the activity of the supernatant, the use of filters is risky; first the absence of lipase adsorption on the filter should be confirmed. Centrifugation may be a good alternative to filtration.

TIP 3.6

If immobilization is slow; it may be due to the existence of lipase dimers that cannot become immobilized on this support. To accelerate the immobilization rate, dilute the sample or add some mild solvents (e.g., 20–50%, v/v; glycerin, PEG).

6. STEP 3. COVALENT IMMOBILIZATION OF ADSORBED LIPASES ON OCTYL–GLYOXYL AGAROSE

6.1 Overview

The lipase molecules are already immobilized via their open form on the octyl–glyoxyl support, but this is no guarantee for the promotion of covalent attachments between enzyme and support (Rueda et al., 2015). To increase the reactivity of the amino groups in lateral chain of Lys, the use of pH 10 is recommended. In some instances, a lower pH like 9 may be enough to get at least one enzyme–support imine bond formation (Rueda et al., 2015). The layer of octyl groups is over the layer of glyoxyl groups, making the enzyme–support covalent reaction difficult, and may be that only lipase groups located around the active center of the enzyme may participate in the reaction (Rueda et al., 2016). After reduction of the imine bonds to secondary amino ones, these bonds become irreversible and the lipase will not be released to the medium under experimental conditions, and this approach may be used to check if really there is at least one covalent bond between enzyme and support.

Duration

2–25 h.

Protocol

4.1. Incubate the enzyme immobilized on octyl–glyoxyl agarose beads in sodium bicarbonate buffer at pH 10 and 25 °C.

4.2. Take samples periodically, reduce these samples with sodium borohydride (1 mg/mL) and incubate them in a detergent solution at a concentration that releases all enzyme molecules adsorbed on octyl–agarose via interfacial activation, using gentle stirring.

4.3. When all enzyme remains in the support after incubation with detergent, we can consider that all enzyme molecules have at least one enzyme–support covalent bond.

4.4 In some cases, a certain percentage of lipase molecules remains desorbed when incubated with detergents and this percentage did not decrease when the treatment time is prolonged. In these cases, we should decide if we want to keep this noncovalently attached enzyme subpopulation or if we prefer to eliminate it by washing with a detergent.

4.5. Reduce the imino groups of the biocatalyst with sodium borohydride (1 mg/mL added in solid form to the immobilization suspension) by 30 min of incubation.

4.6. Wash the reduced preparation with a buffer at pH 7 to remove unreacted sodium borohydride and then wash with buffer or water for storage.

4.7. A SDS-PAGE of the supernatant obtained after boiling the biocatalyst in breaking buffer will confirm the covalent attachment of the enzyme lipase molecules (the secondary amino groups are stable enough for this treatment, thus if at least a covalent bond between the enzyme and the support has been achieved, the lipase band should not be visualized).

TIP 4.1

Lipase stability at pH 10 may be problematic. Incubate an octyl–lipase preparation at pH 10 and check the activity overtime. If there is a significant activity decrease in 24 h, use 4 °C assay additives to improve enzyme stability (e.g., saccharides and other polyols). Avoid the use of any aminated compound during the incubation step (glycine, Tris, etc.).

TIP 4.2

Thiolated compounds are known to stabilize imino bonds. Thus, the addition of thiolated compounds may be utilized to form glyoxyl–lipase covalent attachment at lower pH (Bolivar et al., 2009).

TIP 4.3

Stirring is only necessary when a sample is taken to measure the enzyme activity, now all enzyme molecules are already immobilized, the reaction is intermolecular and pH changes are minimum.

TIP 4.4

1 mg/mL sodium borohydride is not enough to significantly affect a standard protein but some proteins may be more sensitive to this reagent. If this is the case, the reduction may be performed in the presence of nonaminated enzyme protective reagents, or at 4 °C. If still there is a significant loss of activity, the reductant may be used at lower concentrations, just ensuring that a molar excess is used to ensure that all imine and aldehyde groups are reduced. For example, using 1 g of support, 1 mg/mL is recommended if there are 5 mL of total suspension volume, but may be reduced to 0.1 mg/mL using 50 mL.

TIP 4.5

The reduction needs to be performed in a fumehood; otherwise the hydrogen released may produce problems (e.g., breaking of the recipient and loss of suspension when opened).

TIP 4.6

Leave the immobilized enzyme in an aqueous solution containing some antimicrobial agent (e.g., sodium azide) to prevent microbial contamination, and also to prevent excessive drying of the support, if it is going to be stored for weeks.

ACKNOWLEDGMENTS
We gratefully recognize the support from the MINECO of Spanish Government, CTQ2013-41507-R. The predoctoral fellowships for Ms. Rueda (Colciencias, Colombian Government) and Mr. dos Santos (CNPq, Brazil) are also recognized. The authors wish to thank Mr. Ramiro Martínez (Novozymes, Spain) for kindly supplying the enzymes used in this research. The help and comments from Dr. Ángel Berenguer (Instituto de Materiales, Universidad de Alicante) are kindly acknowledged.

REFERENCES
Barbosa, O., Torres, R., Ortiz, C., Berenguer-Murcia, A., Rodrigues, R. C., & Fernandez-Lafuente, R. (2013). Heterofunctional supports in enzyme immobilization: From traditional immobilization protocols to opportunities in tuning enzyme properties. *Biomacromolecules*, 14, 2433–2462.

Bernal, C., Illanes, A., & Wilson, L. (2014). Heterofunctional hydrophilic-hydrophobic porous silica as support for multipoint covalent immobilization of lipases: Application to lactulose palmitate synthesis. *Langmuir*, 30, 3557–3566.

Bolivar, J. M., López-Gallego, F., Godoy, C., Rodrigues, D. S., Rodrigues, R. C., Batalla, P., et al. (2009). The presence of thiolated compounds allows the immobilization of enzymes on glyoxyl agarose at mild pH values: New strategies of stabilization by multipoint covalent attachment. *Enzyme and Microbial Technology*, 45, 477–483.

Fernandez-Lafuente, R., Armisén, P., Sabuquillo, P., Fernández-Lorente, G., & Guisán, J. M. (1998). Immobilization of lipases by selective adsorption on hydrophobic supports. *Chemistry and Physics of Lipids*, 93, 185–197.

Fernandez-Lorente, G., Filice, M., Lopez-Vela, D., Pizarro, C., Wilson, L., Betancor, L., et al. (2010). Cross-linking of lipases adsorbed on hydrophobic supports: Highly selective hydrolysis of fish oil catalyzed by RML. *Journal of American Oil Chemists' Society*, 88, 801–807.

Grochulski, P., Li, Y., Schrag, J. D., Bouthillier, F., Smith, P., Harrison, D., et al. (1993). Insights into interfacial activation from an open structure of *Candida rugosa* lipase. *Journal of Biological Chemistry*, 268, 12843–12847.

Jaeger, K. E., & Eggert, T. (2002). Lipases for biotechnology. *Current Opinion in Biotechnology*, 13, 390–397.

Jaeger, K. E., & Reetz, M. T. (1998). Microbial lipases form versatile tools for biotechnology. *Trends in Biotechnology*, 16, 396–403.

Manoel, E. A., dos Santos, J. C. S., Freire, D. M. G., Rueda, N., & Fernandez-Lafuente, R. (2015). Immobilization of lipases on hydrophobic supports involves the open form of the enzyme. *Enzyme and Microbial Technology*, 71, 53–57.

Mateo, C., Abian, O., Bernedo, M., Cuenca, E., Fuentes, M., Fernandez-Lorente, G., et al. (2005). Some special features of glyoxyl supports to immobilize proteins. *Enzyme and Microbial Technology*, 37, 456–462.

Palomo, J. M., Fuentes, M., Fernández-Lorente, G., Mateo, C., Guisan, J. M., & Fernández-Lafuente, R. (2003). General trend of lipase to self-assemble giving bimolecular aggregates greatly modifies the enzyme functionality. *Biomacromolecules*, 4, 1–6.

Palomo, J. M., Muñoz, G., Fernández-Lorente, G., Mateo, C., Fernández-Lafuente, R., & Guisán, J. M. (2002). Interfacial adsorption of lipases on very hydrophobic support (octadecyl–Sepabeads): Immobilization, hyperactivation and stabilization of the open form of lipases. *Journal of Molecular Catalysis B: Enzymatic, 19–20,* 279–286.

Rueda, N., dos Santos, J. C. S., Ortiz, C., Barbosa, O., Fernandez-Lafuente, R., & Torres, R. (2016). Chemical amination of lipases improves their immobilization on octyl-glyoxyl agarose beads. *Catalysis Today. 259, 107–118.* http://dx.doi.org/10.1016/j.cattod.2015.05.027.

Rueda, N., dos Santos, J. C. S., Torres, R., Ortiz, C., Barbosa, O., & Fernandez-Lafuente, R. (2015). Improved performance of lipases immobilized on heterofunctional octyl-glyoxyl agarose beads. *Royal Society of Chemistry Advances, 5,* 11212–11222.

Schmid, R. D., & Verger, R. (1998). Lipases: Interfacial enzymes with attractive applications. *Angewandte Chemie International Edition, 37,* 1608–1633.

Suescun, A., Rueda, N., dos Santos, J. C. S., Castillo, J. J., Ortiz, C., Torres, R., et al. (2015). Immobilization of lipases on glyoxyl–octyl supports: Improved stability and reactivation strategies. *Process Biochemistry, 50,* 1211–1217.

Verger, R. (1997). "Interfacial activation" of lipases: Facts and artifacts. *Trends in Biotechnology, 15,* 32–38.

Winkler, F. K., D'Arcy, A., & Hunziker, W. (1990). Structure of human pancreatic lipase. *Nature, 343,* 771–774.

CHAPTER FIVE

Biomimetic/Bioinspired Design of Enzyme@capsule Nano/Microsystems

J. Shi*,†, Y. Jiang‡, S. Zhang§, D. Yang*,†, Z. Jiang†,§,1

*School of Environmental Science and Engineering, Tianjin University, Tianjin, China
†Collaborative Innovation Center of Chemical Science and Engineering (Tianjin), Tianjin, China
‡School of Chemical Engineering and Technology, HeBei University of Technology, Tianjin, China
§Key Laboratory for Green Chemical Technology of Ministry of Education, School of Chemical Engineering and Technology, Tianjin University, Tianjin, China
1Corresponding author: e-mail address: zhyjiang@tju.edu.cn

Contents

1. Introduction — 88
    1.1 Introduction of Enzyme@capsule Nano/Microsystems — 88
    1.2 The State-of-the-Art Methods for the Design and Construction of Enzyme@capsule Nano/Microsystems — 89
    1.3 Advantages of Incorporating Biomimetic/Bioinspired Chemistries in the Design and Construction of Enzyme@capsule Nano/Microsystems — 90
2. General Procedure of the Design and Construction of Enzyme@capsule Nano/Microsystems Through Biomimetic/Bioinspired Methods — 93
    2.1 Generation of Enzyme@$CaCO_3$ Templates Through Co-precipitation of Protein Containing $CaCl_2$ and Na_2CO_3 — 93
    2.2 Surface Coating on the Templates Through Biomimetic/Bioinspired Methods — 95
    2.3 Removal of the $CaCO_3$ Components Through EDTA or Dilute HCl Treatment — 95
    2.4 General Characterizations of Enzyme@capsule Nano/Microsystems — 95
3. Some Specific Examples — 97
    3.1 Biomimetic/Bioinspired Mineralization Approach for the Synthesis of Enzyme@capsule Nano/Microsystems — 97
    3.2 Biomimetic/Bioinspired Adhesion (Catechol Chemistry and Polyphenol Chemistry) Approach for the Synthesis of Enzyme@capsule Nano/Microsystems — 103
    3.3 Combined Biomimetic/Bioinspired Mineralization and Adhesion Approach for the Synthesis of Enzyme@capsule Nano/Microsystems — 106
4. Concluding Remarks — 108
Acknowledgment — 110
References — 110

Abstract

Enzyme@capsule nano/microsystems, which refer to the enzyme-immobilized capsules, have received tremendous interest owing to the combination of the high catalytic activities of encapsulated enzymes and the hierarchical structure of the capsule. The preparation of capsules and simultaneous encapsulation of enzymes is recognized as the core process for the rational design and construction of enzyme@capsule nano/microsystems. The strategy used has three major steps: (a) generation of the templates, (b) surface coating on the templates, and (c) removal of the templates, and it has been proven to be effective and versatile for the construction of enzyme@capsule nano/microsystems. Several conventional methods, including layer-by-layer assembly of polyelectrolytes, liquid crystalline templating method, etc., were used to design and construct enzyme@capsule nano/microsystems, but these have two major drawbacks. One is the low mechanical stability of the systems and the second is the harsh conditions used in the construction process. Learning from nature, several biomimetic/bioinspired methods such as biomineralization, biomimetic/bioinspired adhesion, and their combination have been exploited for the construction of enzyme@capsule nano/microsystems. In this chapter, we will present a general protocol for the construction of enzyme@capsule nano/microsystems using the latter approach. Some suggestions for improved design, construction, and characterization will also be presented with detailed procedures for specific examples.

1. INTRODUCTION

1.1 Introduction of Enzyme@capsule Nano/Microsystems

As one kind of fascinating catalysts in nature, enzymes can catalyze numerous transformations with high activity and regio/stereoselectivity, and this aspect of enzymes has been successfully applied in diverse areas, including clean energy products, pharmaceutical/fine/commodity chemicals, etc. (Benkovic & Hammes-Schiffer, 2003; Bornscheuer et al., 2012; Drauz, Groger, & May, 2012; Gross, Kumar, & Kalra, 2001; Shi, Jiang, et al., 2015). To ensure the high stabilities and facile recycling of enzymes, a variety of methods have been developed and tentatively utilized for enzyme immobilization, primarily through strategies such as surface grafting, covalent cross-linking, encapsulation, intercalation, and entrapment (Bornscheuer, 2003; Deshapriya & Kumar, 2013; Hanefeld, Gardossi, & Magner, 2009; Luckarift, Spain, Naik, & Stone, 2004; Rodrigues, Ortiz, Berenguer-Murcia, Torres, & Fernández-Lafuente, 2013; Sheldon, 2007). In this connection, diverse kinds of scaffolds are being made for enzyme immobilization and these include but not limited to nanoparticles, capsules, gels, films, foams, 2D materials, etc.

Among the existing enzyme immobilization supports, capsules are particularly attractive due to their hierarchical structures, upon which "enzyme@capsule nano/microsystems" or "enzyme-encapsulated capsules" can be constructed (Chen, Ye, Zhou, & Wu, 2013; He, Cui, & Li, 2009; Shi, Jiang, et al., 2014; Shi, Shen, & Möhwald, 2004; Tong, Song, & Gao, 2012; Van Dongen et al., 2009). Specifically, in the "enzyme@capsule nano/microsystems," capsules can maintain the encapsulated enzyme in the free state, effectively protected from leakage/environmental attacks, while enhancing enzyme stability. Besides, the pore size and porosity of the capsule walls can be controlled to allow the substrates/products to diffuse between the bulk solution and the interior of the capsules. Owing to the above merits, major efforts are being made to design and construct enzyme@capsule nano/microsystems for advanced applications.

1.2 The State-of-the-Art Methods for the Design and Construction of Enzyme@capsule Nano/Microsystems

In the past decades, three methods have been developed for the construction of enzyme@capsule nano/microsystems and these include template-free, soft-templating, and hard-templating strategies. By contrast, hard-templating strategy is popular owing to several advantages of these templates such as their uniform particle size, easily regulated structure, and controllable synthesis process (Caruso, Trau, Möhwald, & Renneberg, 2000; Kreft, Prevot, Möhwald, & Sukhorukov, 2007; Ortac et al., 2014; Price, Zelikin, Wang, & Caruso, 2009; Van Gough, Wolosiuk, & Braun, 2009).

One of the pioneering methods is the layer-by-layer (LbL) assembly of oppositely charged polyelectrolytes on the surface of enzyme microcrystals. However, two drawbacks were often noticed: the difficulty in controlling the enzyme crystallization process and the mechanical weakness of the polyelectrolyte capsule wall. In most cases, only one of the two drawbacks is addressed. For instance, instead of using enzyme crystals as the templates, Caruso and coworkers (Price et al., 2009) preadsorbed enzymes in mesoporous silica microspheres and then capped them with multilayered polyelectrolytes. After removing the silica template surrounding the enzyme molecules, the enzyme@capsule nano/microsystems were adopted for conducting the catalytic reactions. Braun and coworkers (Van Gough et al., 2009) also adopted silica microspheres as the templates as enzyme-loading support on which an ordered ZnS coating was formed through a lyotropic liquid crystalline templating. After dissolution of the silica templates by treatment with hydrofluoric acid, enzyme@capsule nano/microsystems were obtained. Owing to the ordered pore structure of the capsule wall, this nano/microsystem showed size-selective feature for substrates and can be used for different catalytic purposes.

A more biocompatible way of entrapping enzymes is to coprecipitate enzymes during the formation of water-insoluble carbonates, and the carbonate template can be removed in a much milder way, for example, by using EDTA or dilute HCl solutions. Several research groups have utilized $CaCO_3$ as the template for enzyme support and constructed enzyme@capsule nano/microsystems. Particularly, Kreft's and Gao's groups (Kreft et al., 2007; Tong et al., 2012) both conducted LbL assembly of polyelectrolytes on the presynthesized enzyme@$CaCO_3$ microspheres. After simply immersing the encapsulated enzyme/carbonate product into EDTA or dilute HCl solutions, the water-insoluble carbonates are removed, leaving behind the enzyme@capsule nano/microsystems. However, the mechanical stability of these capsules is still a problem because of their organic components of the capsule walls. In short, the enzyme@capsule nano/microsystems developed so far have some drawbacks, either during the construction of the capsule or in the subsequent utilization of the encapsulated enzyme. Use of harsh conditions such as *extreme high/low pH value, or toxic reagents, and others* in the encapsulation process can denature or destabilize the delicate enzyme or the low mechanical stability of the capsule walls, which can pose problems during the catalytic applications of the capsules.

Therefore, the strategy needs to improved to overcome these issues, and much more attention should be paid to the construction of the capsule wall under milder conditions that are more compatible with the delicate enzymes being used. Also, the mechanical stability of the capsule walls is to be significantly enhanced so that the biocatalysts can withstand harsh reaction conditions and for recycling, so that the cost of the biocatalyst could be lowered.

1.3 Advantages of Incorporating Biomimetic/Bioinspired Chemistries in the Design and Construction of Enzyme@capsule Nano/Microsystems

In nature, organisms use a number of amazing strategies to synthesize high-performance materials under extremely mild conditions such as neutral pH, room temperature, and aqueous solution conditions. This may inspire us to learn and master some powerful techniques from nature for the rational design and facile construction of enzyme@capsule nano/microsystems. Among them, biomineralization is a well-known process by which living organisms assemble hierarchical nano/microstructures from naturally occurring inorganic compounds to generate biominerals for hardening or stiffening existing tissues. During the biomineralization process, inorganic ions in the surrounding regions would first aggregate and precipitate on the preassembled organics (mainly referring to proteins/polyelectrolytes) and then form biominerals under the influence of the organic substances. Up to date, nearly seventy kinds of biominerals have been identified, some of which are rather familiar to us, e.g., the silica in diatoms, the calcium phosphate in the bones and teeth of the vertebrates, and the calcium carbonate matrix in the nacre (Gower, 2008).

Usually, the materials synthesized through biomineralization have more delicate structure, and consequently, they have superior physicochemical properties when compared to the materials derived from conventional methods. Additionally, the physiological environment in living organisms dictates that the biominerals be synthesized under mild conditions. For instance, biosilicification in diatoms is an exciting example (Hildebrand, 2008; Vrieling, Beelen, van Santen, & Gieskes, 1999). In diatoms, the silica deposition vesicle is considered as the cellular "reaction vessel," where all the chemical steps of silica formation and patterning would take place. The organic constituents (proteins and long-chain polyamine) in diatom cells have been recognized to play important roles in mediating biosilicification process. Specifically, the diatom cells can control the assembly of these organic constituents on their cell membrane and then utilize the assembled organic constituents to induce the silification process. The formed organic/biosilica shell that surrounded on the cell membrane can confer the diatom with high mechanical stability against the external environment (Roth, Zhou, Yang, & Morse, 2005). Thus, formation of a biocompatible matrix under mild conditions is a biological lesson worth examining for the capsule formation.

Besides biomineralization, bioadhesion is another intriguing phenomenon found in organisms, which demonstrate how natural substances adhere to a solid surface in a rapid and robust way. In particular, marine organisms have set fascinating examples of bioadhesion which is commonly found in marine organisms, such as mussels, sandcastle worms, tube worms, limpets, and sea cucumbers. Typically, marine mussels can secrete adhesive proteins, which can adhere to a solid surface and harden shortly to generate a solidified layer in water so that the mussels can be firmly attached on nearly all types of substrates including rocks and ship hulls. Through intensive study, *Mytilus edulis* foot protein 3 (Mefp-3) and Mefp-5 are identified as important for contributing to the mussels' adhesive capacity (Lee, Dellatore, Miller, & Messersmith, 2007; Lee, Scherer, & Messersmith, 2006). Both Mefp-3 and Mefp-5 are known to have a high content of 3,4-dihydroxy-L-phenylalanine (DOPA), which plays a critical role in the adhesion. For example, DOPA is known to: (1) participate in the reactions leading to the hardening of bulk adhesive proteins and (2) form strong covalent and noncovalent interactions with solid substrates due to the chemically diverse functionalities of the catechol groups on DOPA.

Additionally, metal ions are shown to be essential during the bioadhesive process (Wilker, 2010). Especially, iron–DOPA complexes in the byssus of mussel contribute much to the improvement of the hardness and extensibility of the threads. Such bioadhesives have some distinct advantages, of which the mild and rapid formation process in aqueous phase which is also desired in the enzyme encapsulation strategies. To mimic this process, tremendous effort has been devoted to screen or synthesize analogs of bioadhesives. For example, dopamine, which was discovered by Messersmith's group (Lee et al., 2007; Lee, Messersmith, Israelachvili, & Waite, 2011), has been widely utilized as an adhesive mimic due to its structure/property similarity to DOPA. Plant polyphenol has also been discovered as another adhesive mimic by his group and shown great impact on a diverse set of research areas. This burgeoning area of bioadhesion will undoubtedly help in the rational design and synthesis of multifunctional nanomaterials.

As described above, nature can implement both biomineralization and bioadhesion processes to manufacture high-performance materials under ambient conditions. Introducing these nature-inspired methods into the design and construction of enzyme@capsule nano/microsystem may provide a novel, rational way to pursue both goals of achieving mild process for the construction of the capsule and high mechanical stability of the capsule wall.

Typically, nature-inspired methods refer to biomimetic/bioinspired methods. To the best of our knowledge, the term of "biomimetics" was put forward in the 1960s. This area was defined as the study of the structure–function relationships of biological systems in the rational design of engineering solutions. So, biomimetics usually means directly imitating or copying natural processes or structures, which may or may not be the most suitable solutions to the problem at hand. There is no doubt that biomimetics can play an important role in exploratory research, but extensive implementation of biomimetics to technological fields is yet to be fully realized.

Such transition of fundamental science to technological implementation is just the field that bioinspiration takes over. The term of "bioinspiration" was introduced in the past decade, and it emphasizes the extraction of fundamental ideas/principles from biological systems and the utilization of these ideas/principles in the construction of high-performance materials/structures with little or no resemblance to their corresponding biological prototypes. Obviously, biomimetics and bioinspiration cover a variety of disciplines, including biological science, material science, chemistry, chemical engineering, and so on. In this context, the idea of biomimetic/bioinspired rational design and construction of enzyme@capsule nano/microsystem is anticipated to have implications for a broad range of scientific areas. These concepts are implemented in the following methods of the enzyme@capsule formation and their characterization.

2. GENERAL PROCEDURE OF THE DESIGN AND CONSTRUCTION OF ENZYME@CAPSULE NANO/MICROSYSTEMS THROUGH BIOMIMETIC/BIOINSPIRED METHODS

The general construction process of enzyme@capsule nano/microsystems can be depicted as shown in Fig. 1. Detailed procedure for its implementation is given below.

2.1 Generation of Enzyme@CaCO₃ Templates Through Co-precipitation of Protein Containing CaCl₂ and Na₂CO₃

The enzyme@$CaCO_3$ microspheres are prepared and utilized as templates for the preparation of enzyme capsules (Jiang et al., 2009; Kreft et al., 2007; Petrov, Volodkin, & Sukhorukov, 2005; Zhang et al., 2014). Briefly,

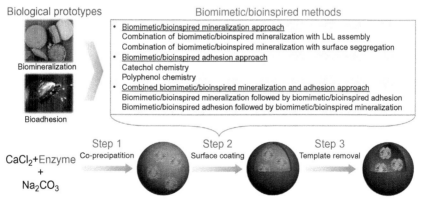

Figure 1 Rational design of enzyme@capsule nano/microsystems via biomimetic/bioinspired methods.

Na$_2$CO$_3$ aqueous solution (a certain concentration, no higher than 330 mM, pH 10.9) is rapidly poured into an equal volume of an aqueous solution of CaCl$_2$ (same concentration as Na$_2$CO$_3$ solution, pH 6.3), polyelectrolyte (a certain amount, no higher than 3 mg mL^{-1}), and enzymes (a certain amount, no higher than 2 mg mL^{-1}, pH 7.0) under vigorous stirring for 30 s. After stationary settling for a certain time (no longer than 30 min), enzyme@CaCO$_3$ microspheres are obtained through centrifugation (3000 rpm, 3 min) and washing with deionized water. Herein, the polyelectrolyte is utilized for stabilizing the crystalline phase of CaCO$_3$ and rendering the particle with a charged surface. The particle size of the enzyme@CaCO$_3$ microspheres can be regulated by changing the stirring rate during the mixing process and by adjusting the concentration of CaCl$_2$ and/or Na$_2$CO$_3$ solutions.

Higher stirring rate usually results in smaller particle sizes. The enzyme loading (w/w) increases gradually with increasing enzyme concentration, which then reaches a constant value at a certain enzyme concentration. Excess enzyme loading may lower the enzymatic efficiency. Therefore, an appropriate enzyme concentration is usually pre-evaluated before conducting the encapsulation process. Moreover, the concentrations of CaCl$_2$ or Na$_2$CO$_3$ solutions should not be higher than 330 mM, and the settling time during the mixing process should be shorter than 30 min. Otherwise, the initial crystalline form of the solid may change as a result of recrystallization to other forms.

Due to the delicate nature of enzymes and their sensitivity to solution conditions, several aspects should be taken into consideration during the formation process of enzyme@CaCO$_3$ microspheres:

(1) *Charge (pI) of enzymes*: Enzymes with a lower isoelectric point (pI value, probable <5.0) may directly coprecipitate with Ca^{2+} ions at neutral pH and inhibit the formation of enzyme@CaCO$_3$ templates.

(2) *Size of enzymes*: Considering that the diameters of most enzymes are less than 20 nm, which is much smaller than that of the CaCO$_3$ templates formed, the size of enzymes may have little or no influence on the encapsulation process.

(3) *Hydrophobicity of enzymes*: Enzymes with a high hydrophobicity index may not dissolve in the aqueous phase, and therefore, they are difficult to be entrapped in the inorganic solid matrix.

(4) *Side chains of enzymes*: Enzymes that carry significant numbers of particular side chains, such as histidines which are known to have a high affinity for Ca^{2+}, may be unsuitable. The effect is similar to that of enzyme

charge, since the enzyme may coordinate with significant numbers of Ca^{2+} ions prior to the addition of $NaCO_3$, alter its charge, and adversely influence the formation of enzyme@$CaCO_3$ templates.

2.2 Surface Coating on the Templates Through Biomimetic/Bioinspired Methods

The enzyme@$CaCO_3$ microspheres described above are dispersed into a solution that contains precursors for coating or construction of the capsule walls. After the coating, the microspheres can be acquired through centrifugation (3000 rpm, 3 min) and washing with deionized water. Although the coating procedures are different for different biomimetic/bioinspired methods, they are all conducted under mild conditions (neutral pH, room temperature, and aqueous solution). Generally, most of the examples can be summarized and categorized into three types: (1) coating synthesized through biomimetic/bioinspired mineralization, (2) coating synthesized through biomimetic/bioinspired adhesion, and (3) coatings synthesized through the combination of the above. Detailed synthesis procedures, structure regulation and cautions are given in Section 3.

2.3 Removal of the $CaCO_3$ Components Through EDTA or Dilute HCl Treatment

Enzyme@capsule nano/microsystems are obtained after removing the templates through EDTA (15 mM, pH 6.5–7.0) or dilute HCl treatment. The pH of the template removal conditions should be above 6.0, because lower pH may lead to an irreversible damage to the enzyme structure resulting in substantial loss of its activity. After the removal of the templates, rinsing with water is required before further use or characterization of the capsules.

2.4 General Characterizations of Enzyme@capsule Nano/Microsystems

2.4.1 Verification of the Formation Process of Enzyme@capsule Nano/Microsystems

Since the formation of enzyme@capsule nano/microsystems involves three typical steps, monitoring of each step is essential for the acquirement of the final nano/microsystems. The most important variation to be monitored is the topological revolution, which can be characterized by scanning electron microscopy (SEM) and transmission electron microscopy (TEM). There are several tips for these two characterizations. First of all, the sample should be dispersed in water at a low concentration. The solution is then dropped onto a glass slide for SEM or onto a copper grid for TEM. Second, the wall thickness and hollow structure of the capsule are commonly characterized by TEM, while the large-scale view and surface morphology of the capsules are generally examined by SEM. However, in some cases, the wall thickness for some capsules is too thick to be observed by TEM. Samples with broken capsules generated by grinding are more favorable to examine the wall thickness through cross-sectional SEM images. Third, all samples need not be coated or stained prior to imaging for SEM or TEM studies.

Another issue to monitor is the change in chemical/physical structures, which can be characterized by Fourier transform-infrared spectroscopy (FTIR), X-ray photoelectron spectroscopy (XPS), and Brunner–Emmet–Teller (BET) measurements. For the FTIR studies, the samples should be fully dried or else the strong absorption by water can mask the details of the expected transitions from the samples. For the BET analysis, the degassing temperature could be 90 °C or lower to inhibit the shrinkage and collapse of the capsule structure. XPS is widely used for quantitative measurement of the elemental composition of the surface (<10 nm thickness) of a material. The synthesis of the enzyme@capsule nano/microsystems by the above method contains a surface-coating process, after which the surface elemental composition would be altered. This change is quantitatively detected and analyzed by XPS. Energy dispersive X-ray spectroscopy, an accessory to the SEM, although less accurate than XPS, is an alternative technique to measure the elemental composition of the walls of the enzyme@capsule nano/microsystems.

2.4.2 Location of the Enzyme in Enzyme@capsule Nano/Microsystems

The acquirement of information about the precise location of enzyme in the enzyme@capsule nano/microsystems is important. Nowadays, the only reliable characterization technique is confocal laser scanning microscopy (CLSM). The preparation of the samples for this method is very complicated. One should first stain the enzymes with dyes such as fluorescein isothiocyanate, followed by a weeklong dialysis to obtain the dye-linked enzymes. Notably, during the dialysis process, water or buffer is to be changed multiple times. Otherwise, a lot of the free dye would be left in the enzyme@capsule nano/microsystems, which would affect the CLSM result. Similar to the method of making samples for SEM/TEM, the as-constructed enzyme@capsule nano/microsystems should be dropped onto a glass slide at low concentrations and covered with the cover glass slips for CLSM studies.

2.4.3 Activity/Stability of Enzyme@capsule Nano/Microsystems

The catalytic activity and stability are two key properties for evaluating the performance of enzyme@capsule nano/microsystems. Accordingly, several specific parameters, including conversion ratio, productivity, Michaelis constant (K_m), maximum velocity of the reaction (V_{max}), and pH/storage/thermal/recycling stabilities, are to be determined. Amongst these, the conversion ratio, productivity, K_m and V_{max} are utilized to reflect the enzyme catalytic activity, whereas the pH/storage/thermal/recycling stabilities are measured to reflect on the enzyme stability. Different enzymes involve different species substrates and products, which would determine the choice of methods for monitoring the catalytic activities. And this method could involve ultraviolet–visible spectroscopy, gas chromatography, high-performance liquid chromatography, etc.

In the following sections, several specific examples will be illustrated, which mainly focus on the descriptions of the design and construction procedures of enzyme@capsule nano/microsystems. Some cautions/notes are also mentioned (also see Table 1).

3. SOME SPECIFIC EXAMPLES

3.1 Biomimetic/Bioinspired Mineralization Approach for the Synthesis of Enzyme@capsule Nano/Microsystems

3.1.1 Combination of Biomimetic/Bioinspired Mineralization with LbL Assembly (Jiang et al., 2009; Li et al., 2010; Shi, Zhang, & Jiang, 2011)

The protamine–titania capsules are used in this method to produce the corresponding enzyme@capsule nano/microsystems (denoted as System I, enzymes used in this method are alcohol dehydrogenase, formate dehydrogenase, and catalase) by combining the biomimetic/bioinspired mineralization with the LbL assembly (Jiang et al., 2009).

3.1.1.1 Detailed Synthesis Procedure

During the construction of System I, poly(sodium-p-styrenesulfonate) (PSS, 70,000 kDa)-doped enzyme@$CaCO_3$ microspheres are used as templates, which are prepared according to the co-precipitation method previously described in Section 2.1.

The PSS-doped enzyme@$CaCO_3$ microspheres (25 mg) are then dispersed in protamine aqueous solution (5 mL, 2 mg mL^{-1}, pH 7.0)

Table 1 Summary of the Key Points for the Six Specific Examples/Cases

Specific Methods		Key Points	Enzymes (Used in These Methods)	Net Efficiency of Enzyme Encapsulation (in lumen)	Activity Retained After (n) Cycles
Section 3.1: Biomimetic/bioinspired mineralization	Section 3.1.1: Combination of biomimetic/bioinspired mineralization with LbL assembly	• The protamine layer dual roles: (1) acts as the positively charged layer electrostatically attracting the negatively charged Ti-BALDH layer, (2) acts as the biocatalyst inducing the in situ conversion of Ti-BALDH into titania. • Protamine can be replaced by other cationic proteins or polymers. • Titania precursor can be substituted with other inorganic precursors (such as acid).	Alcohol dehydrogenase, formate dehydrogenase, catalase	99.0%	62.0% of initial activity retained after six cycles
	Section 3.1.2: Combination of biomimetic/bioinspired mineralization with surface segregation	• PAH has two critical roles: (1) serves as the hydrophilic polymer rendering the segregating capacity and spontaneous enrichment in the near-surface region of $CaCO_3$ microspheres,	Formate dehydrogenase, catalase	99%	98.0% of initial activity retained after 10 cycles

		(2) serves as the cationic polymer offering the confined space and inducing the formation of titania minerals. • The $CaCO_3$ microspheres act as dual templates for forming both capsule lumen and mesopores on the capsule wall. • GA can be replaced by other cross-linkers (such as tannic acid (TA), dopamine). • Titania precursor can be substituted with other inorganic precursors (such as silicic acid).		
Section 3.2: Biomimetic/bioinspired adhesion	Section 3.2.1: Catechol chemistry	• The self-oxidative polymerization of dopamine makes the central contribution to forming an intact layer on the surface of the templates. • Dopamine can be replaced by other catecholamines (such as norepinephrine, DOPA).	α-Glucosidase, catalase	35.0%	24.0% of initial activity retained after seven cycles
	Section 3.2.2: Polyphenol chemistry	• The strong interfacial affinity of TA is responsible for forming the polyphenol coating through oxidative oligomerization.	Glucose oxidase	91.0%	30.0% of initial activity retained after 16 cycles

Continued

Table 1 Summary of the Key Points for the Six Specific Examples/Cases—cont'd

Specific Methods		Key Points	Enzymes (Used in These Methods)	Net Efficiency of Enzyme Encapsulation (in lumen)	Activity Retained After (n) Cycles
		• The high reactivity of TA is in charge of reacting/cross-linking with cationic polymer PEI through Schiff base/Michael addition reaction. • TA can be replaced by other polyphenols (such as epigallocatechin gallate (EGCG), oligomeric proanthocyanidins (OPC)). • PEI can be substituted with other cationic polymers.			
Section 3.3: Combined biomimetic/bioinspired mineralization and adhesion	Section 3.3.1: Biomimetic/bioinspired mineralization followed by biomimetic/bioinspired adhesion	• Protamine can be replaced by other cationic polymers. • Titania precursor can be substituted with other inorganic precursors (such as silicic acid).	α-Glucosidase catalase	69%	85.0–100% of initial activity retained after eight cycles
	Section 3.3.2: Biomimetic/bioinspired adhesion followed by biomimetic/bioinspired mineralization	• Cysteamine acts as a bridge between the polydopamine and titania layers to enhance the interfacial stability: (1) the –SH groups of cysteamine can react	Catalase	93.0%	28.0% of initial activity retained after seven cycles

with the catechol groups of polydopamine to form covalent bonds through Michael addition/Schiff base reaction,

(2) the –NH$_2$ groups of cysteamine can induce the formation of titania minerals.

- Cysteamine can be replaced by other cationic molecules or polymers (such as PEI, PAH).
- Dopamine can be replaced by polyphenols.
- Titania precursor can be substituted with other inorganic precursors (such as silicic acid).

containing NaCl (500 mM). After shaking for 15 min, enzyme@CaCO$_3$ microspheres that are coated with protamine layer by electrostatic interactions are collected by centrifugation (3000 rpm, 3 min) and washed with deionized water to remove loosely bound protamine. Subsequently, these microspheres are suspended in titanium (IV) bis(ammonium lactato) dihydroxide (Ti-BALDH) solution (5 mL, 1.25 wt%) and shaken for 15 min. After centrifugation (3000 rpm, 3 min) and washing with deionized water, enzyme@CaCO$_3$ microspheres coated with protamine and a titania layers are obtained. This procedure is repeated until a desired number of protamine/titania layers are deposited on the microcapsules.

As EDTA could chelate Ca^{2+} dissolves CaCO$_3$, System I was finally obtained after removing CaCO$_3$ matrix by EDTA treatment (15 mM, pH 6.0–7.0, incubation time 10 min). The net efficiency of enzyme encapsulation can reach 99.0%.

3.1.1.2 Some Key Points

During the surface-coating process, the protamine layer has dual roles: (1) it may serve as a positively charged layer electrostatically attracting the negatively charged Ti-BALDH layer and (2) serve as the biocatalyst inducing the *in situ* conversion of Ti-BALDH into titania.

In this method, protamine can be replaced by other cationic proteins or polymers, while titania precursor can be substituted with other inorganic precursors such as silicic acid to build the corresponding inorganic coating.

3.1.2 Combination of Biomimetic/Bioinspired Mineralization with Surface Seggregation (Shi, Zhang, Wang, & Jiang, 2014; Shi, Zhang, et al., 2013)

In this method, the enzyme@capsule nano/microsystem (denoted as System II, enzymes used in this method are formate dehydrogenase and catalase) is constructed based on mesoporous hybrid capsules, which is prepared by combining biomimetic/bioinspired mineralization and surface seggregation (Shi, Zhang, et al., 2013).

3.1.2.1 Detailed Synthesis Procedure

During the construction of System II, polyallylamine hydrochloride (PAH, 70,000 kDa)-doped enzyme@CaCO$_3$ (P-enzyme@CaCO$_3$) microspheres are used as the templates, which are prepared according to the co-precipitation method previously described in Section 2.1.

Then, P-enzyme@$CaCO_3$ microspheres (5 mg) are dispersed in 1.25 mL of glutaric dialdehyde (GA) aqueous solution (pH 6.8 ± 0.2). The GA concentrations are tuned from 0.01 to 2.00 wt%. After keeping in a vessel under gentle agitation for a certain period of time (10–240 min), the centrifugation (3000 rpm, 3 min) and washing with deionized water are conducted for several times until excess GA is washed away.

Next, PAH/GA-enzyme@$CaCO_3$ (PG-enzyme@$CaCO_3$) microspheres dispersed in 1 mL of Ti-BALDH aqueous solution (5–100 mM, pH 7.0), the mixture was maintained in a vessel under gentle agitation for 60 min, centrifuged (3000 rpm, 3 min), and washed several times with deionized water until the excess Ti-BALDH has been washed away.

Finally, PAH/GA-titania-enzyme@$CaCO_3$ (PGTi-enzyme@$CaCO_3$) microspheres were incubated in 1 mL, EDTA solution (15 mM, pH 6.5–7.0) for 5 min under shaking to obtain System II, washed with deionized water till the wash was free of EDTA. The net efficiency of enzyme encapsulation can reach 99%.

3.1.2.2 Some Key Points

During the surface-coating process, PAH has two critical roles: (1) it serves as the hydrophilic polymer rendering the seggregating capacity and spontaneous enrichment in the near-surface region of $CaCO_3$ microspheres and (2) functions as the cationic polymer offering a confined space for the formation of titania minerals. The $CaCO_3$ microspheres act as the dual templates for forming both capsule lumen and mesopores on the capsule wall.

In this method, GA can be replaced by other cross-linkers such as tannic acid (TA), or dopamine, while titania precursor can be substituted with other inorganic precursors such as silicic acid.

3.2 Biomimetic/Bioinspired Adhesion (Catechol Chemistry and Polyphenol Chemistry) Approach for the Synthesis of Enzyme@capsule Nano/Microsystems

3.2.1 Catechol Chemistry (Shi, Yang, et al., 2013; Zhang, Shi, Jiang, Jiang, Qiao, et al., 2011)

Inspired by the structural organization of mitochondria and the bioadhesive principles, the enzyme@capsule nano/microsystem (denoted as System III, enzymes used in this method are α-glucosidase and catalase) based on polydopamine capsules is constructed through catechol chemistry (Zhang, Shi, Jiang, Jiang, Qiao, et al., 2011).

3.2.1.1 Detailed Synthesis Procedure

During the construction of System III, PSS-doped enzyme@$CaCO_3$ microspheres are used as the templates, which are prepared according to the co-precipitation method previously described in Section 2.1.

The enzyme@$CaCO_3$ microspheres (50 mg) are prewashed with Tris–HCl buffer (100 mM, pH 7.5–8.5) through several rounds of centrifugation (3000 rpm, 3 min) followed by redispersion cycles. The resulting enzyme@$CaCO_3$ microspheres in Tris–HCl buffer (10 mL, 100 mM, pH 8.5) were treated with dopamine hydrochloride (2–10 mg mL^{-1}). The suspension was allowed to equilibrate for 2–10 h, with constant shaking. The resulting tan-colored particles are centrifuged (3000 rpm, 3 min) and washed with Tris–HCl buffer until the supernatant was colorless.

As EDTA could chelate Ca^{2+} dissolve $CaCO_3$, System III was obtained after the complete removal of the $CaCO_3$ by treatment with the EDTA solution (15 mM, pH 6.5–7.0, incubation time 10 min). The net efficiency of enzyme encapsulation can reach 35.0%.

3.2.1.2 Some Key Points

During the synthesis, the self-oxidative polymerization of dopamine makes important contribution to forming a cross-linked layer on the surface of the template. Although the polymerization mechanism is still under investigation, dopamine would be first oxidized into 5,6-dihydroxyindole (DHI), and physical aggregation of DHI (noncovalent bonds) as well as the chemical cross-linking of DHI (covalent bonds) contribute to the generation of polydopamine coating on the capsule surface.

In this method, dopamine can be replaced by other catecholamines (such as norepinephrine or DOPA).

3.2.1.3 Extension of This Method to the Construction of Multienzyme Systems

The above methods were also extended to construct multienzyme containing microcapsule systems. One advantage of multienzyme systems is that a cascade of reactions can be carried out by the enzyme mixture, when the enzymes are appropriately chosen for the cascade chemistry. Similarly, the multienzyme systems can be used to carry out different reactions with the same capsules, when different enzymes are chosen, without having to change the catalyst.

To immobilize three enzymes by polydopamine capsules, enzyme I was added to prepare PSS-doped $CaCO_3$ microspheres according to the procedure described earlier. Then, the acquired (enzyme I)@$CaCO_3$

microspheres were used as the templates. These $CaCO_3$ microspheres were resuspended in Tris–HCl (100 mM, pH 7.5–8.5) solution containing enzyme II and dopamine for 2–10 h with constant shaking. Next, the resulting tan-colored particles were centrifuged (3000 rpm, 3 min) and washed three times with fresh Tris–HCl buffer. Finally, the resultant microspheres were resuspended in enzyme III solution for 8 h with constant shaking. The decomposition of the $CaCO_3$ components by EDTA treatment (15 mM, pH 6.5–7.0) was performed as described above. (Example: enzyme I, α-glucosidase, net efficiency of enzyme encapsulation 33.0%; enzyme II, α-amylase, net efficiency of enzyme encapsulation 35.0%; enzyme III, β-amylase, net efficiency of enzyme encapsulation 20.0%.)

In this process, the three enzymes are, respectively, immobilized through physical encapsulation in the lumen (enzyme I), *in situ* entrapment within the capsule wall (enzyme II), and by chemical attachment on the outer surface (enzyme III) (Zhang, Shi, Jiang, Jiang, Qiao, et al., 2011).

3.2.2 Polyphenol Chemistry (Zhang, Jiang, Wang, Yang, & Shi, 2015)

In this method, the enzyme@capsule nano/microsystem (denoted as System IV, enzyme used in this method was glucose oxidase) of polyphenol capsules was constructed through polyphenol chemistry (Zhang, Jiang, Wang, et al., 2015).

3.2.2.1 Detailed Synthesis Procedure

During the construction of System IV, PSS-doped enzyme@$CaCO_3$ microspheres were used as the templates, prepared by the co-precipitation method described in Section 2.1.

The PSS-doped enzyme@$CaCO_3$ microspheres (50 mg) were dispersed in 10 mL Tris–HCl buffer (50 mM, pH 7.5–8.5) containing 0.5 mg mL^{-1} TA. After gentle stirring at room temperature for 2 h, the microspheres were collected by centrifugation (3000 rpm, 3 min) and washing with deionized water several times. The microspheres were redispersed in 10 mL Tris–HCl buffer (50 mM, pH 8.0) containing 1.0 mg mL^{-1} polyethylenimine (PEI, 1800, 3600, 70,000 kDa) and stirred at room temperature for 0.5–2 h. The microspheres were collected by centrifugation (3000 rpm, 3 min) followed by washing with water several times.

System IV was obtained after removing the $CaCO_3$ components through EDTA treatment (15 mM, pH 6.5–7.0). The net efficiency of enzyme encapsulation can reach 91.0%.

3.2.2.2 Some Key Points

During the surface-coating process, the strong interfacial affinity of TA was responsible for forming the polyphenol coating through oxidative oligomerization, whereas the high reactivity of TA resulted in reacting/cross-linking with the cationic PEI through Schiff base/Michael addition reaction.

In this method, TA can be replaced by other polyphenols (such as epigallocatechin gallate (EGCG), oligomeric proanthocyanidins (OPC)), while PEI can be substituted with other cationic polymers.

3.3 Combined Biomimetic/Bioinspired Mineralization and Adhesion Approach for the Synthesis of Enzyme@capsule Nano/Microsystems

3.3.1 Biomimetic/Bioinspired Mineralization Followed by Biomimetic/Bioinspired Adhesion (Zhang, Shi, Jiang, Jiang, Meng, et al., 2011)

In this method, the enzyme@capsule nano/microsystem (System V, enzyme used in this method was catalase) of protamine/titania/polydopamine capsules was constructed through biomimetic/bioinspired mineralization followed by biomimetic/bioinspired adhesion (Zhang, Shi, Jiang, Jiang, Meng, et al., 2011).

3.3.1.1 Detailed Synthesis Procedure

During the construction of System V, PSS-doped enzyme@$CaCO_3$ microspheres were used as the templates, prepared according to the co-precipitation method described in Section 2.1.

Subsequently, the PSS-doped enzyme@$CaCO_3$ microspheres (25 mg) were dispersed in protamine aqueous solution (5 mL, 2 g L^{-1}) containing NaCl (500 mM). After shaking for 15 min, enzyme@$CaCO_3$ microspheres coated with protamine were collected by centrifugation (3000 rpm, 3 min) and washed twice with water to remove the residual protamine. These microspheres were suspended in Ti-BALDH solution (5 mL, 1.25 wt%), shaken for 15 min, centrifuged (3000 rpm, 3 min), and then washed with deionized water to obtain the $CaCO_3$ microspheres coated with protamine and titania. The resultant microspheres are gathered and resuspended in 20 mL of Tris–HCl buffer (100 mM, pH 8.5) with a concentration of 2 mg mL^{-1} of dopamine hydrochloride, and the suspension is allowed to proceed for 8 h with constant shaking. Next, the tan-colored particles are centrifuged (3000 rpm, 3 min) and washed with fresh Tris–HCl buffer until the supernatant became colorless.

System V was obtained after the removal of the CaCO$_3$ components through incubating the microspheres in EDTA solution (15 mM, pH 6.5–7.0). The net efficiency of enzyme encapsulation can reach 69%.

3.3.1.2 Some Key Points
For this method, protamine can be replaced by other cationic polymers, while titania precursor can be substituted with other inorganic precursors.

3.3.1.3 Extension of This Method to the Construction of Multienzyme System (Zhang, Shi, Jiang, Jiang, Meng, et al., 2011)
This method was also extended to construct multienzyme systems. To immobilize three enzymes by the protamine/titania/polydopamine microcapsules, 1 mL of enzyme I was added during the co-precipitation process to prepare the PSS-doped CaCO$_3$ microspheres. As the templates, enzyme I@CaCO$_3$ microspheres were first dispersed in protamine aqueous solution (5 mL, 2 g L^{-1}) containing NaCl (500 mM), shaken for 15 min, collected by centrifugation (3000 rpm, 3 min), and washed with deionized water to remove the residual protamine. Then, these microspheres were suspended in 5 mL of 1.25 wt% Ti-BALDH solution, shaken for 15 min, centrifuged (3000 rpm, 3 min), and washed with water. The microspheres were resuspended in Tris–HCl (100 mM, pH 7.5–8.5) containing enzyme II and dopamine for 2–10 h with constant shaking, centrifuged (3000 rpm, 3 min), and washed with fresh Tris–HCl buffer. Finally, the resultant microspheres were resuspended in enzyme III solution for 8 h with constant shaking at room temperature. The decomposition of the CaCO$_3$ components by EDTA treatment (15 mM, pH 6.5–7.0) was performed as previously described. (Example: enzyme I, α-glucosidase, net efficiency of enzyme encapsulation 69%; enzyme II, α-amylase, net efficiency of enzyme encapsulation 47%; enzyme III, β-amylase, net efficiency of enzyme encapsulation 50%.)

3.3.2 Biomimetic/Bioinspired Adhesion Followed by Biomimetic/Bioinspired Mineralization (Shi, Zhang, Zhang, Wang, & Jiang, 2015; Zhang, Jiang, Zhang, Wang, & Shi, 2015; Zhang et al., 2014)
In this method, the enzyme@capsule nano/microsystem (System VI, enzymes used in this method were α-glucosidase and catalase) on the basis of protamine/cysteamine/titania capsules was constructed via biomimetic/bioinspired adhesion followed by biomimetic/bioinspired mineralization (Zhang et al., 2014).

3.3.2.1 Detailed Synthesis Procedure

In the synthesis of System VI, PSS-doped enzyme@$CaCO_3$ microspheres were used as the templates, as prepared according to the co-precipitation method previously described in Section 2.1.

The PSS-doped enzyme@$CaCO_3$ microspheres (25 mg) were suspended in 5 mL of Tris–HCl buffer (100 mM, pH 7.5–8.5), then 10 mg of dopamine hydrochloride has been added. The mixture was kept under stirring for 8 h at room temperature, centrifuged (3000 rpm, 3 min), and washed with Tris–HCl buffer until the supernatant became colorless. Thus, enzyme@$CaCO_3$ microspheres coated with a polydopamine layer were obtained. Subsequently, the microspheres were resuspended in 5 mL of Tris–HCl buffer in the presence of 16 mg mL^{-1} cysteamine, and the suspension kept under stirring for 2 h at room temperature. Excess cysteamine was removed by centrifugation (3000 rpm, 3 min) and washed with deionized water. The microspheres were suspended in a Ti-BALDH solution (5 mL, 1.25 wt%) to coat the titania layer. After stirring for 2 h, the resulting microspheres were collected by centrifugation (3000 rpm, 3 min) and washed with deionized water to remove the residual Ti-BALDH.

The $CaCO_3$ components were removed through incubating the microspheres with EDTA solution (15 mM, pH 6.5–7.0). System VI is obtained after centrifugation (3000 rpm, 3 min) and washing with deionized water. The net efficiency of enzyme encapsulation can reach 93.0%.

3.3.2.2 Some Key Points

During the surface-coating process, cysteamine acts as a bridge between the polydopamine and titania layers to enhance the interfacial stability. The –SH groups of cysteamine react with the catechol groups of polydopamine to form covalent bonds through Michael addition/Schiff base reaction, and the –NH_2 groups of cysteamine induce the formation of titania minerals.

In this method, cysteamine can be replaced by other cationic molecules or polymers (such as PEI, PAH); dopamine can be replaced by polyphenols; and titania precursor can be substituted with other inorganic precursors.

4. CONCLUDING REMARKS

In summary, enzyme@capsule nano/microsystems have proven to be efficient biocatalysts and successfully applied in a variety of catalytic processes. Enzymes confer reactivity to capsules, while capsules provide appropriate microenvironments for the enzymes. Although tremendous effort has been made to design and construct enzyme@capsule nano/microsystems, biomimetic/bioinspired methods provide an alternative and fascinating way to deal with this issue under mild conditions. Among, two typical biomimetic/bioinspired methods, biomimetic/bioinspired mineralization and biomimetic/bioinspired adhesion, are most frequently utilized. And the as-constructed nano/microsystems show comparable or even higher performance than other systems derived from conventional methods. Nonetheless, some aspects should be further examined in the future, which may allow the biomimetic/bioinspired methods to evolve into a much more generic, reliable, and versatile platform technologies. Several key points may be summarized:

1. In this chapter, we have illustrated the methods of biomimetic/bioinspired mineralization, biomimetic/bioinspired adhesion, and their combination, which are three major methods for the construction of enzyme@capsule nano/microsystems. More biological prototypes will be discovered in the near future, which can offer alternative or more efficient methods for the design and construction of enzyme@capsule nano/microsystems.

2. The performance of an enzyme@capsule nano/microsystem is not only affected by its construction method but also by its hierarchical structure. Since the existing investigations mainly lay stress on exploiting and screening of suitable methods, further research should pay attention to the detailed and quantitative investigation of the structure–catalytic performance relationships of the nano/microsystems.

3. We should gradually switch our focus from the fundamental research to industrialization. This would help to push the concept into practical application, such as biosensing, pharmaceutical synthesis, and fine chemicals production.

4. To date, biomimetic/bioinspired methods have been utilized to construct enzyme@capsule nano/microsystems for enzyme immobilization. As another important research branch of biocatalysis, the whole-cell catalysis shares much similarity to enzyme catalysis as the immobilization process is essentially required. Extension of the biomimetic/bioinspired methods to immobilize cells may bring unexpected benefits for the whole-cell catalysis.

5. Natural enzymes/whole cells are capable of catalyzing a broad variety of reactions for different applications. Nonetheless, there are still a lot of useful chemicals that cannot be produced through naturally occurring enzymatic/whole-cell routes. Fortunately, artificial enzymes, which are

designed and synthesized via synthetic biology, are being evolved as an essential supplement to natural enzymes. The biomimetic/bioinspired methods introduced here could also be applied to immobilize these artificial enzymes for further constructing the corresponding enzyme@capsule nano/microsystems.

ACKNOWLEDGMENT
The authors thank the financial support from the National Science Fund for Distinguished Young Scholars (21125627), National Natural Science Funds of China (21406163), Tianjin Research Program of Application Foundation and Advanced Technology (15JCQNJC10000), and Program of Introducing Talents of Discipline to Universities (B06006).

REFERENCES
Benkovic, S. J., & Hammes-Schiffer, S. (2003). A perspective on enzyme catalysis. *Science, 301*, 1196–1202.
Bornscheuer, U. T. (2003). Immobilizing enzymes: How to create more suitable biocatalysts. *Angewandte Chemie, International Edition, 42*, 3336–3337.
Bornscheuer, U. T., Huisman, G. W., Kazlauskas, R. J., Lutz, S., Moore, J. C., & Robins, K. (2012). Engineering the third wave of biocatalysis. *Nature, 485*, 185–194.
Caruso, F., Trau, D., Möhwald, H., & Renneberg, R. (2000). Enzyme encapsulation in layer-by-layer engineered polymer multilayer capsules. *Langmuir, 16*, 1485–1488.
Chen, M., Ye, C., Zhou, S., & Wu, L. (2013). Recent advances in applications and performance of inorganic hollow spheres in devices. *Advanced Materials, 25*, 5343–5351.
Deshapriya, I., & Kumar, C. V. (2013). Nano-Bio interfaces: Charge control of enzyme/inorganic interfaces for advanced biocatalysis. *Langmuir, 29*, 14001–14016.
Drauz, K., Groger, H., & May, O. (2012). *Enzyme catalysis in organic synthesis: A comprehensive handbook* (3rd ed.). Weinheim, Germany: John Wiley & Sons.
Gower, L. B. (2008). Biomimetic model systems for investigating the amorphous precursor pathway and its role in biomineralization. *Chemical Reviews, 108*, 4551–4627.
Gross, R. A., Kumar, A., & Kalra, B. (2001). Polymer synthesis by in vitro enzyme catalysis. *Chemical Reviews, 101*, 2097–2124.
Hanefeld, U., Gardossi, L., & Magner, E. (2009). Understanding enzyme immobilisation. *Chemical Society Reviews, 38*, 453–468.
He, Q., Cui, Y., & Li, J. (2009). Molecular assembly and application of biomimetic microcapsules. *Chemical Society Reviews, 38*, 2292–2303.
Hildebrand, M. (2008). Diatoms, biomineralization processes, and genomics. *Chemical Reviews, 108*, 4855–4874.
Jiang, Y., Yang, D., Zhang, L., Sun, Q., Sun, X., Li, J., et al. (2009). Preparation of protamine-titania microcapsules through synergy between layer-by-layer assembly and biomimetic mineralization. *Advanced Functional Materials, 19*, 150–156.
Kreft, O., Prevot, M., Möhwald, H., & Sukhorukov, G. B. (2007). Shell-in-shell microcapsules: A novel tool for integrated, spatially confined enzymatic reactions. *Angewandte Chemie, International Edition, 46*, 5605–5608.
Lee, H., Dellatore, S. M., Miller, W. M., & Messersmith, P. B. (2007). Mussel-inspired surface chemistry for multifunctional coatings. *Science, 318*, 426–430.
Lee, B. P., Messersmith, P. B., Israelachvili, J. N., & Waite, J. H. (2011). Mussel-inspired adhesives and coatings. *Annual Review of Materials Research, 41*, 99.

Lee, H., Scherer, N. F., & Messersmith, P. B. (2006). Single-molecule mechanics of mussel adhesion. *Proceedings of the National Academy of Sciences of the United States of America, 103*, 12999–13003.

Li, J., Jiang, Z., Wu, H., Zhang, L., Long, L., & Jiang, Y. (2010). Constructing inorganic shell onto LbL microcapsule through biomimetic mineralization: A novel and facile method for fabrication of microbioreactors. *Soft Matter, 6*, 542–550.

Luckarift, H. R., Spain, J. C., Naik, R. R., & Stone, M. O. (2004). Enzyme immobilization in a biomimetic silica support. *Nature Biotechnology, 22*, 211–213.

Ortac, I., Simberg, D., Yeh, Y. S., Yang, J., Messmer, B., Trogler, W. C., et al. (2014). Dual-porosity hollow nanoparticles for the immunoprotection and delivery of non-human enzymes. *Nano Letters, 14*, 3023–3032.

Petrov, A. I., Volodkin, D. V., & Sukhorukov, G. B. (2005). Protein-calcium carbonate coprecipitation: A tool for protein encapsulation. *Biotechnology Progress, 21*, 918–925.

Price, A. D., Zelikin, A. N., Wang, Y., & Caruso, F. (2009). Triggered enzymatic degradation of DNA within selectively permeable polymer capsule microreactors. *Angewandte Chemie, International Edition, 48*, 329–332.

Rodrigues, R. C., Ortiz, C., Berenguer-Murcia, A., Torres, R., & Fernández-Lafuente, R. (2013). Modifying enzyme activity and selectivity by immobilization. *Chemical Society Reviews, 42*, 6290–6307.

Roth, K. M., Zhou, Y., Yang, W. J., & Morse, D. E. (2005). Bifunctional small molecules are biomimetic catalysts for silica synthesis at neutral pH. *Journal of the American Chemical Society, 127*, 325–330.

Sheldon, R. A. (2007). Enzyme immobilization: The quest for optimum performance. *Advanced Synthesis and Catalysis, 349*, 1289–1307.

Shi, J., Jiang, Y., Jiang, Z., Wang, X., Wang, X., Zhang, S., et al. (2015). Enzymatic conversion of carbon dioxide. *Chemical Society Reviews, 44*, 5981–6000.

Shi, J., Jiang, Y., Wang, X., Wu, H., Yang, D., Pan, F., et al. (2014). Design and synthesis of organic-inorganic hybrid capsules for biotechnological applications. *Chemical Society Reviews, 43*, 5192–5210.

Shi, X., Shen, M., & Möhwald, H. (2004). Polyelectrolyte multilayer nanoreactors toward the synthesis of diverse nanostructured materials. *Progress in Polymer Science, 29*, 987–1019.

Shi, J., Yang, C., Zhang, S., Wang, X., Jiang, Z., Zhang, W., et al. (2013). Polydopamine microcapsules with different wall structures prepared by a template-mediated method for enzyme immobilization. *ACS Applied Materials & Interfaces, 5*, 9991–9997.

Shi, J., Zhang, L., & Jiang, Z. (2011). Facile construction of multicompartment multienzyme system through layer-by-layer self-assembly and biomimetic mineralization. *ACS Applied Materials & Interfaces, 3*, 881–889.

Shi, J., Zhang, S., Wang, X., & Jiang, Z. (2014). Open-mouthed hybrid microcapsules with elevated enzyme loading and enhanced catalytic activity. *Chemical Communications, 50*, 12500–12503.

Shi, J., Zhang, W., Wang, X., Jiang, Z., Zhang, S., Zhang, X., et al. (2013). Exploring the segregating and mineralization-inducing capacities of cationic hydrophilic polymers for preparation of robust, multifunctional mesoporous hybrid microcapsules. *ACS Applied Materials & Interfaces, 5*, 5174–5185.

Shi, J., Zhang, W., Zhang, S., Wang, X., & Jiang, Z. (2015). Synthesis of organic–inorganic hybrid microcapsules through in situ generation of an inorganic layer on an adhesive layer with mineralization-inducing capability. *Journal of Materials Chemistry B, 3*, 465–474.

Tong, W., Song, X., & Gao, C. (2012). Layer-by-layer assembly of microcapsules and their biomedical applications. *Chemical Society Reviews, 41*, 6103–6124.

Van Dongen, S. F., de Hoog, H. P. M., Peters, R. J., Nallani, M., Nolte, R. J., & van Hest, J. C. (2009). Biohybrid polymer capsules. *Chemical Reviews, 109*, 6212–6274.

Van Gough, D., Wolosiuk, A., & Braun, P. V. (2009). Mesoporous ZnS nanorattles: Programmed size selected access to encapsulated enzymes. *Nano Letters, 9*, 1994–1998.

Vrieling, E. G., Beelen, T. P., van Santen, R. A., & Gieskes, W. W. (1999). Diatom silicon biomineralization as an inspirational source of new approaches to silica production. *Journal of Biotechnology, 70*, 39–51.

Wilker, J. J. (2010). The iron-fortified adhesive system of marine mussels. *Angewandte Chemie, International Edition, 49*, 8076–8078.

Zhang, S., Jiang, Z., Wang, X., Yang, C., & Shi, J. (2015). Facile method to prepare microcapsules inspired by polyphenol chemistry for efficient enzyme immobilization. *ACS Applied Materials & Interfaces, 7*, 19570–19578.

Zhang, S., Jiang, Z., Zhang, W., Wang, X., & Shi, J. (2015). Polymer-inorganic microcapsules fabricated by combining biomimetic adhesion and bioinspired mineralization and their use for catalase immobilization. *Biochemical Engineering Journal, 93*, 281–288.

Zhang, L., Shi, J., Jiang, Z., Jiang, Y., Meng, R., Zhu, Y., et al. (2011). Facile preparation of robust microcapsules by manipulating metal-coordination interaction between biomineral layer and bioadhesive layer. *ACS Applied Materials & Interfaces, 3*, 597–605.

Zhang, L., Shi, J., Jiang, Z., Jiang, Y., Qiao, S. Z., Li, J., et al. (2011). Bioinspired preparation of polydopamine microcapsule for multienzyme system construction. *Green Chemistry, 13*, 300–306.

Zhang, W., Shi, J., Wang, X., Jiang, Z., Song, X., & Ai, Q. (2014). Conferring an adhesion layer with mineralization-inducing capabilities for preparing organic-inorganic hybrid microcapsules. *Journal of Materials Chemistry B, 2*, 1371–1378.

CHAPTER SIX

Synergistic Functions of Enzymes Bound to Semiconducting Layers

K. Kamada*,[1], A. Yamada*, M. Kamiuchi*, M. Tokunaga*, D. Ito*, N. Soh[†]

*Graduate School of Engineering, Nagasaki University, Nagasaki, Japan
[†]Faculty of Agriculture, Saga University, Saga, Japan
[1]Corresponding author: e-mail address: kkamada@nagasaki-u.ac.jp

Contents

1. Introduction   114
2. Fabrication of Enzyme-Intercalated Layered Oxides   116
  2.1 Synthetic Procedure of Colloidal Solution of Exfoliated Titanate Layers with Micrometric Dimensions   116
  2.2 Synthesis of Colloidal Solution of Layered Titanate with Nanometric Dimensions   118
  2.3 Binding of Enzymes to Layered Titanate via Physical Interaction   120
3. Activity of Enzymes Bound to Titanate Layers   123
  3.1 Activity of Enzymes Intercalated into Micrometric or Nanometric Titanate Layers   123
  3.2 Anti-UV Light Stability of Enzymes Inserted into Layered Titanates   125
4. Photochemical Control of Enzymatic Activity of Oxidoreductases Bound to Layered Oxides   125
  4.1 Activity Control of Peroxidases Intercalated into FT Layers by UV Light Irradiation   126
  4.2 Visible-Light-Driven Enzymatic Activity of HRP Bound to Semiconductors with a Narrow Band Gap   127
5. Biorecognition Using Doped Titanate Layers Modified with Biomolecules   129
6. Magnetic Application of Hybrids Composed of Enzymes and Doped Titanates   130
  6.1 Hybrids Composed of Enzyme and FT Layers   130
  6.2 Nanohybrids Composed of Enzyme, Titanate Nanosheet, and Magnetic Beads   131
7. Conclusions   132
Acknowledgments   133
References   133

Abstract

Synthesis and cooperative functions of hybrid materials composed of enzyme and semiconducting layers are described in this chapter. The hybrids were produced via a simple physical interaction between the components, that is, electrostatic interaction in an aqueous solution. To form interstratifying enzymes in the galleries, solution pH, which is a key parameter to decide surface potential, should be adjusted appropriately. In other words, enzymes should have an opposite charge when compared to that of the layers at an identical pH. Even though the intercalation slightly reduced enzymatic activity as compared to those of the free enzymes, stability under cruel conditions was drastically improved due to screening effect of semiconducting layers from extrinsic stimuli. In addition, photochemical control of redox enzymes sandwiched between semiconducting layers was accomplished. Light irradiation of the hybrids induced band gap excitation of the layers, and holes produced in the valence band activated the enzymes. It was revealed that the semiconducting layers with magnetic elements might be useful to magnetic application (separation) of enzymes as similar to conventional magnetic beads.

1. INTRODUCTION

As well known, enzymes are biocatalysts existing in organisms that operate even in moderate environments (Zhang, Ge, & Liu, 2015). Their excellent reaction rates and substrate selectivity have been utilized for practical application such as food industries and chemical synthesis of useful substances. Since enzymes originated from living bodies, the use under cruel conditions, for example, at high temperature, in nonphysiological pH, and an organic solvent, is severely restricted (Mozhev et al., 1989). These conditions induce irreversible denaturation corresponding to deformation of three-dimensional framework of an enzyme. To solve the problem, many researchers and manufacturers have attempted to fabricate immobilized enzymes for several decades. Enzymes are fixed to surface of (or inside) a physicochemically stable support (e.g., polymers, inorganic materials) via chemical or physical interaction. Interaction force (binding energy) to a support resulted in stability enhancement (Ikemoto, Chi, & Ulstrup, 2010). In addition, the fact that enzymes with tiny dimensions are bound to a bulky support, realizing the possibility of recovery and reuse of enzymes that are generally expensive and application in flow-type reactors by incorporating a reaction column containing the solid-bound enzymes.

Recently, inorganic layered materials have attracted much attention as novel supports for enzymes (Kumar & Chaudhari, 2000). In this case, enzymes are interstratified with host layers. Most of layered oxides used have a periodical crystal structure that is composed of anionic layers and interlayer cations. Substitution of enzymes for the interlayer cations realizes formation of the new type of immobilized enzyme, that is, nanohybrids of enzymes and oxide layers as schematically illustrated in Fig. 1. There are a few cases that cationic layered hydroxides like layered double hydroxides (LDHs) also impregnate anionic enzymes (Zhang, Chen, Wang, & Yang, 2008). The immobilization of enzyme in the interlayer space is mainly accomplished through electrostatic interaction. The nanohybrid materials consisting of enzymes and layered solids possess many advantages on the basis of the unique structure. Enzymes in the nanohybrids are not immobilized on surface of a support (e.g., fixation on silica gels), nor are covered with a stable shell (incorporation into polymer chains). Intercalation into layers enhances stability of enzyme, and hence accessibility to external substrates remains along the two-dimensional interlayer space. In other words, enzymes are imperfectly covered with oxide layers. Enzymes intercalated have a superior stability at high temperature (Kumar & Chaudhari, 2002) and in organic solvents (Wang, Gao, & Shi, 2004), suggesting that layered materials are useful as hosts for a novel form among various immobilized enzymes.

As similar to other layered solids, layered metal oxides show a variety of functions according to chemical composition that cannot be achieved for biomolecules. They have high physicochemical stability as compared with biomolecules, electrical, magnetic, and optical properties are representative as examples. Therefore, the nanohybrids composed of layered oxides and enzymes are possible to have synergistic functions originated from both components, inducing development of a novel functional material or reaction system. This chapter mentions fabrication and cooperative functions of enzyme-intercalated layered oxides with a semiconducting property.

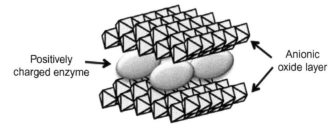

Figure 1 Schematic illustration of enzyme-intercalated oxide layers.

2. FABRICATION OF ENZYME-INTERCALATED LAYERED OXIDES

2.1 Synthetic Procedure of Colloidal Solution of Exfoliated Titanate Layers with Micrometric Dimensions

Colloidal solutions of exfoliated single oxide layers are adequate as raw materials to sandwich large enzyme molecules between thin oxide layers with less than 1 nm in thickness. We can find a number of studies regarding synthesis of colloidal solutions of exfoliated layered metal oxides. In general, parent powder of a layered metal oxide with anionic oxide layers and inlayer cations is prepared by a solid-state reaction at high temperature. Subsequently, the interlayer cations are exchanged for bulky tetraalkylammonium cations in a basic solution to exfoliate to single oxide layers (nanosheets). The obtained colloidal solution is mixed with a solution of the intended enzymes, resulting in formation of nanohybrids between the layers and the enzymes. Here, a fabrication procedure of Fe-doped titanate layers with potassium ions in the interlayer ($K_{0.8}Fe_{0.8}Ti_{1.2}O_4$) is introduced in detail (Harada, Sasaki, Ebina, & Watanabe, 2002).

Commercial powders of K_2CO_3, Fe_2O_3, and TiO_2 in *near* stoichiometry (1.05:1.0:3.0 in molar ratio) were thoroughly mixed in an agate mortar, and then the mixed powder was calcined at 1573 K for 12 h in air. The excess amount of K_2CO_3 (5%) should be added to the raw powder to compensate for the evaporative loss of K_2CO_3. Moreover, the powder must be calcined in a Pt (or other inert) crucible with a Pt cap. If an alumina crucible was used, the K_2CO_3 melt would corrode and then the Al would enter in the powder as an impurity. The resultant powder was reground in the agate mortar followed by recalcination under identical condition to make the $K_{0.8}Fe_{0.8}Ti_{1.2}O_4$ (denoted as KFT) powder with homogeneous elemental distribution. The SEM image of resultant brown powder of KFT is displayed as Fig. 2A. Particles with distinct lamellar structure are clearly observed, and their sizes range from a few to several tens of µm. To stimulate exfoliation to single FT layers ($Fe_{0.8}Ti_{1.2}O_4^{0.8-}$), the interlayer K^+ ions were exchanged for protons by dispersing the KFT powder in an aqueous solution of 1 M HCl (ca. 20 mg/mL) and subsequent stirring for 24 h. Then the solution was replaced with a fresh 1 M HCl, after sedimentation of the solid component. The replacement was repeated several times to complete the ion exchange. The appearance of powder changes to yellowish brown after the proton exchange. Figure 2B shows the SEM image of the protonated FT (HFT),

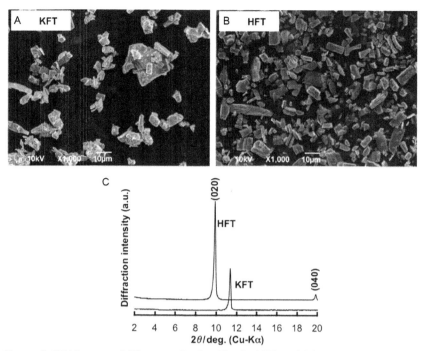

Figure 2 SEM images of (A) parent $K_{0.8}Fe_{0.8}Ti_{1.2}O_4$ (KFT) and (B) protonated powders (HFT). (C) XRD patterns of the KFT and the HFT.

indicating that the proton exchange proceeded with maintaining the particle size and the lamellar microstructure. According to the X-ray diffraction (XRD) patterns of the KFT and the HFT (Fig. 2C), d-spacing of (020) plane perpendicular to two-dimensional FT layers slightly increased by the proton exchange because of larger dimensions of proton (H_3O^+) than K^+.

The HFT powder with the expanded interlayer space was suspended in a tetrabutylammonium (TBA^+) hydroxide solution (4 mg/mL), where concentration of the TBA^+ was adjusted for proton-exchanging capacity of the HFT. The suspension (typically 100 mL) was viscously stirred over 1 week. The protons in the HFT were replaced to the bulky TBA^+, and then the HFT was exfoliated to single FT layers, because the TBA^+ might not attract the FT layers due to its low charge density as compared to the small inorganic monovalent cations. Figure 3A is the photograph of a reactor during the exfoliation. The appearance of the suspension gradually became glossy with an increase in the stirring time, suggesting progress of the exfoliation to single FT layers. When a typical magnetic stir bar coated with

Figure 3 Photographs of colloidal solutions of protonated Fe-doped titanates during (A) exfoliation with a tetrabutylammonium hydroxide and (B) after the exfoliation followed by neutralization and dilution (pH 8.3). (See the color plate.)

PTFE was employed, the surface was peeled off as a result of polishing by titanate nanoparticles, and hence propeller-type stirring or shaking is appropriate for exfoliation. Since addition of the basic colloidal solution would cause denaturation of enzymes, the solution was neutralized to pH 7–8 by titrating with a 1 M acetic acid solution. Reduction of pH less than ca. 6.5 brought about aggregation of the FT layers. Finally, unexfoliated HFT particles were separated by centrifugation and then the transparent and yellowish colloidal solution of FT layers was obtained as shown in Fig. 3B. Concentrations of Ti and Fe in the final solution were estimated by inductively coupled plasma atomic emission (ICP-AE) spectrometry or colorimetry with H_2O_2 after dissolving the FT layers in concentrated H_2SO_4 solution.

2.2 Synthesis of Colloidal Solution of Layered Titanate with Nanometric Dimensions

The aforementioned solid-state reaction and subsequent exfoliation consume a great quantity of thermal energy and time. Hence, we have prepared a colloidal solution of oxide layers with nanometric dimensions through a one-step hydrolysis reaction on the basis of the technique previously reported by Ohya and coworkers with some minor modifications (Ohya, Nakayama, Ban, Ohya, & Takahashi, 2002), and then the produced layers were also employed as supports for enzymes. In the case of titanate, a detailed fabrication procedure was as follows. Liquid titanium(IV) tetraisopropoxide (TTIP) was hydrolyzed by adding an aqueous solution of TBA^+ hydroxide at room temperature. The Ti concentration and the molar ratio of Ti for

TBA$^+$ in the mixture were 0.3 M and unity, respectively. Typically, total volume of the mixture was adjusted to 1–4 mL depending on the amount required. Although the mixed solution was turbid just after mixing, the colorless and transparent colloidal solution clearly scattering a laser beam was produced by shaking at 333 K for 2 h. When doped titanates are prepared, an absolute ethanol solution of dopant metal salt is mixed with TTIP prior to the hydrolysis (Kamada & Soh, 2015). The resultant basic colloidal solution was concentrated by using a centrifugal filter unit (molecular weight cutoff: 3000 Da, centrifugation force: 12,000 $\times g$). The retentate was diluted with deionized water and concentrated again. The process was repeated several times until the solution was nearly neutral (pH 9). During the neutralization, by-products such as isopropanol were also removed from the solution. The yield of Ti in the colloidal solution is estimated to be ca. 30% which was estimated by the ICP spectrometry. Raman spectroscopy revealed that the solid particle in the solution had crystal structure similar to tetratitanate ($Ti_4O_9^{2-}$). Figure 4 depicts a particle size distribution curve of titanate layers doped with Eu^{3+} in the solution after neutralizing. The particle size distribution was monodispersive and a mean size was less than 10 nm. Such tiny (nanometric) size is one of the features of the colloidal solution that is fabricated with the liquid phase synthesis. Besides titanate layers, layered niobates and tantalates can also be prepared when used with an appropriate metal alkoxide (niobium(V) ethoxide, etc.) (Ban, Yoshikawa, & Ohya, 2011).

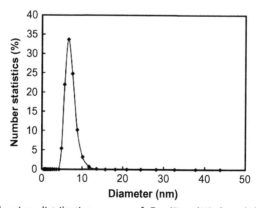

Figure 4 Particle size distribution curve of Eu (5 mol%)-doped layered titanates prepared by a hydrolysis reaction of titanium(IV) tetraisopropoxide with an tetrabutylammonium hydroxide solution.

2.3 Binding of Enzymes to Layered Titanate via Physical Interaction

In order to bind enzyme molecules to the obtained layered titanates, the colloidal solution of titanate layers was mixed with an aqueous enzyme solution. We have tried to bind them through a simple physical interaction, that is, electrostatic interaction. According to zeta potential measurement, the FT layers ($Fe_{0.8}Ti_{1.2}O_4^{0.8-}$) had a negative surface charge over a wide range of pH and its isoelectric point (pI) was ca. 2. The pH dependence of zeta potential of FT layers has been described in our previous manuscript (Kamada, Tsukahara, & Soh, 2010). As a matter of course, a net surface charge of enzyme depends on balance of numbers of anionic and cationic residues (Deshapriya, Kim, Novak, & Kumar, 2014). Consequently, to carry out the binding, one should adjust solution pH to ensure that both materials have opposite surface charges.

An actual binding process is explained as follows. Here, the exfoliated FT layers are intercalated with cellulases that catalyze a hydrolysis of glycosidic bonds by way of example. Figure 5 shows pH dependence of zeta potential of cellulase molecules. Likewise in most proteins, the zeta potential was shifted to a negative direction with an increase in pH and a pI of cellulase was estimated to be ca. 4.7. Hence, it was expected that the FT layers and the cellulase had to be mixed at a pH less than 4.7 to induce electrostatic interaction. Addition of the FT layers (0.74 mg/mL) to a cellulase (0.5 mg/mL) solution buffered at several pHs caused formation of aggregates. These aggregates were collected by centrifugation and then washed with a copious amount of water to remove unreacted species. The XRD

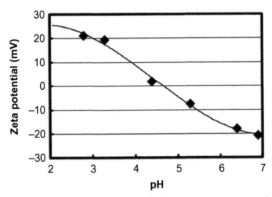

Figure 5 Effect of pH on zeta potential of cellulase in an aqueous solution.

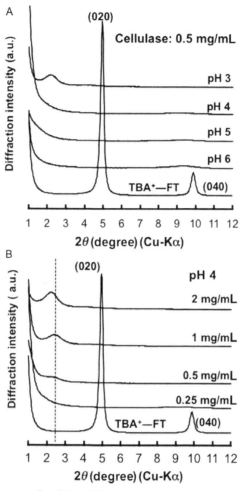

Figure 6 XRD patterns of cellulase-FT composites prepared at several (A) pHs and (B) cellulase concentrations at pH 4. In all processes, concentration of FT was fixed to 0.74 mg/mL.

patterns of the aggregates are displayed in Fig. 6A. The pattern of FT layers after drying the colloidal solution without the addition of cellulase (TBA$^+$-FT) has a main peak assigned to (020) plane of Fe$_{0.8}$Ti$_{1.2}$O$_4^{0.8-}$ layers intercalated with TBA$^+$ ions. The d-spacing of (020) was calculated to be 1.8 nm which is close to the sum of the thickness of FT (0.75 nm) and ion size of TBA$^+$ (0.9 nm). The aggregates did not show any diffraction peak when prepared at pH 4–6, implying that these pHs were inappropriate for the binding. In contrast, the sample fabricated at pH 3 had a broad (020) peak at lower diffraction angle than the TBA^+-FT, suggesting an expansion of the interlayer space due to intercalation of cellulases into the FT layers. In addition, the interlayer distance increased with the concentration of cellulase affecting the intercalation amount (Fig. 6B). As a result, the periodical structure between the FT layers and cellulases as illustrated in Fig. 1 was formed by simple mixing. The result of Fig. 6A indicates that low pH is adequate to induce the strong electrostatic interaction with the FT layers unless the enzyme is denaturated. The SEM image of the precipitates synthesized at pH 3 is shown in Fig. 7, indicating that the layered structure has been maintained even after the intercalation of cellulases. Figure 8 shows weight ratios of cellulases to the FT layers in the precipitates as a function of cellulase concentration in the reactant solution. The ratio linearly increased at the low concentration range, less than 1 mg/mL and was saturated over the range. The saturated binding density ($\sim$0.6 w/w) for the FT layers was largely exceeded those of other supports. For example, a binding density for PMMA was $\sim$0.06 w/w (Ho, Mao, Gu, & Li, 2008). The fact suggests that the exfoliated FT layers have a huge surface area and hence have a large binding capacity for enzymes. Consequently, cellulases were immobilized in the interlayer space via electrostatic interaction at an appropriate pH.

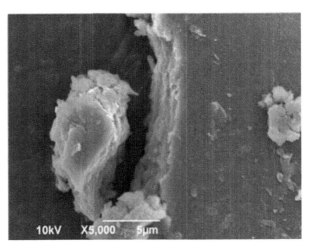

Figure 7 SEM image of cellulase intercalated FT layers (weight ratio of cellulase to FT = ca. 0.6).

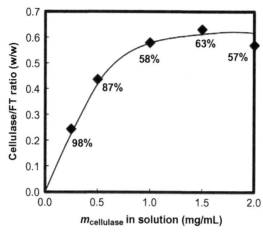

Figure 8 Influence of cellulase concentration in a mixed solution with FT (0.74 mg/mL) on weight ratios of cellulase against FT in precipitates. The percentages in the figure indicate binding efficiencies of cellulase.

3. ACTIVITY OF ENZYMES BOUND TO TITANATE LAYERS

3.1 Activity of Enzymes Intercalated into Micrometric or Nanometric Titanate Layers

Enzymatic activity of cellulase existing in the interlayer of FT was evaluated and compared with free (unbound) cellulase. Scheme 1 depicts a reaction utilized for the activity assessment. Cellulase catalyzes hydrolysis of glycosidic bonds in carboxy methyl cellulose (CMC), resulting in production of glucose. Therefore, quantitative analysis of glucose produced as a result of the enzymatic reaction was carried out in order to evaluate enzymatic activity. Concretely, 3,5-dinitrosalicylic acid (DNSA) was reduced by glucose at 373 K for 1 h, and then 3-amino-5-nitrosalicylic acid was quantified by measuring optical absorption at 530 nm. As expected, the absorbance at 530 nm increased with glucose concentration. Figure 9 compares long-term stabilities of cellulases before (free cellulase) and after intercalation into the FT layers (Cellulase-FT), where initial activity of free cellulase is standardized as 100%. As-prepared cellulase-FT had a slightly smaller activity than free cellulase (82%). This may be due to steric hindrance by the intercalation that inhibits access of large CMC molecules to be hydrolyzed. An apparent activity reduction was also confirmed for horseradish peroxidase (HRP). Figure 9 implies that both cellulases possessed about 80% of activity even after storing for 28 days at 277 K. During the storage,

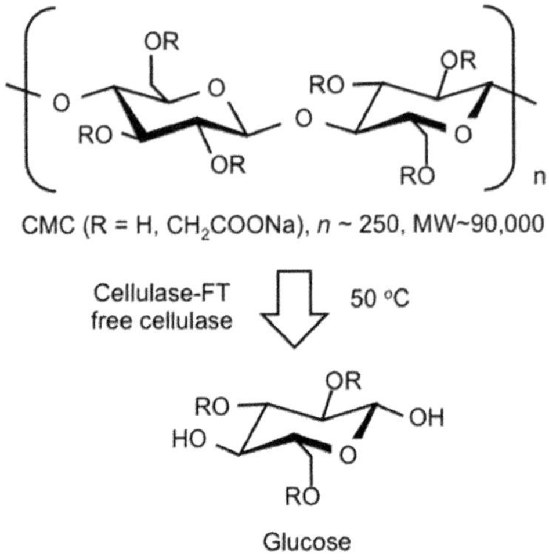

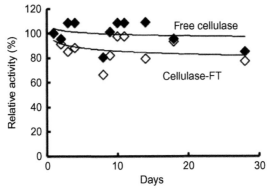

Scheme 1 Reaction scheme of decomposition of carboxy methyl cellulose (CMC) with cellulase-FT or free cellulase.

Figure 9 Enzymatic activity retention of free cellulase (closed diamonds) and FT layers intercalated with cellulase (open diamonds). The concentrations of cellulase and FT were 0.29 and 0.48 mg/mL at pH 6.2, respectively. Activity evaluation was carried out by using decomposition of CMC (0.1 mg/mL) at 323 K and subsequent quantitative analysis of produced glucose with a DNSA method ($n=4$). Between the activity evaluations, the solutions of enzymes were stored at 277 K.

the free or bound cellulase had been dispersed in pure water. The data are average values of four identical assays. At this moment, we have not confirmed activity enhancement of enzymes originated from the KFT synthesized by the solid-state reaction. However, advantages of immobilized enzymes such as facile collection from a reactant solution and reuse are effective for the enzyme-intercalated FT layers.

Recently, we have reported that addition of the nanometric titanate layers prepared by the one-step hydrolysis reaction aforementioned, enhances enzymatic activity of HRP especially at dilute enzyme concentrations and a specific pH (Kamada, Yamada, & Soh, 2015). That is, maximum reaction velocity (V_{max}) in the Michaelis–Menten equation became more than twice by hybridization of HRP with the titanates. The agglomerates composed of several HRP molecules at a diluted solution were dissociated by weak interaction with the nanometric titanate layers, inducing increase of apparent concentration of HRP. Similar activity enhancement was also confirmed for a photoprotein, aequorin (Kamada, 2014).

3.2 Anti-UV Light Stability of Enzymes Inserted into Layered Titanates

The hybridization of enzymes with the FT layers is useful especially under cruel conditions. Improved stability at high temperature or in organic solvents has already been reported. Therefore, our interest was directed to stabilization of HRP against UV light by the intercalation into the FT layers (HRP-FT) because enzymatic activity of HRP-FT could be controlled by UV light irradiation. Solutions including free HRP or the HRP-FT ($w_{HRP}=0.12$ mg/mL, 10 mL) were exposed to UV light from 300 W Hg-Xe lamp (main emission: $\lambda=365$ nm) at room temperature, and then change in enzymatic activity has been investigated (Kamada, Tsukahara, & Soh, 2011). In contrast, the free HRP completely lost catalytic activity after the irradiation for 5 min, the HRP-FT retained 50% of initial activity. Half life of activity was 0.4 and 5.0 min for the free HRP and the HRP-FT, respectively. Typically, an undoped layered titanate ($Ti_{1.825}O_4^{0.7-}$) behaves as a wide band gap n-type semiconductor with an absorption edge at ca. 350 nm. On the other hand, an absorption edge of the FT layers is located at ca. 400 nm, indicating that the FT layers can absorb most UV range. As a result, the HRP sandwiched between the FT layers was protected from UV light of high photon energy.

4. PHOTOCHEMICAL CONTROL OF ENZYMATIC ACTIVITY OF OXIDOREDUCTASES BOUND TO LAYERED OXIDES

The experimental details (protocols) in this section are explained in our previous manuscripts cited.

4.1 Activity Control of Peroxidases Intercalated into FT Layers by UV Light Irradiation

As one example of synergistic function of enzyme–inorganic layer nanohybrids, we have tried to control activity of peroxidase (HRP) intercalated into the FT layers by UV light irradiation (Kamada, Nakamura, & Tsukahara, 2011). Figure 10 illustrates an expected photoreaction model. UV light irradiation causes band gap excitation of the FT layers, and electrons and holes are produced in the conduction (CB) and valence band (VB), respectively. The holes left in the VB oxidize/activate an active center of HRP. The excited HRP (Ox.) oxidizes substrates (Sub.) together with regeneration to the resting state (Red.). That is, the hole is used as an oxidizer instead of H_2O_2 utilized for conventional catalytic oxidation with HRP. Since the holes are produced only under UV light irradiation, the HRP activity will be controlled by irradiation intensity and/or wavelength affecting a hole density in the VB.

To perform the photoinduced enzymatic reaction, UV light (365 nm) was irradiated to the solutions including free HRP or the HRP-FT at room temperature. Amplex UltraRed (AUR) was selected as a substrate. AUR is non-fluorescent but is transformed into a fluorescent molecule by oxidation with HRP (Fruk, Rajendran, Spengler, & Niemeyer, 2007). Therefore, fluorescence intensity of the solution was monitored during the UV light irradiation. Figure 11 indicates changes in relative fluorescence unit (RFU) during intermittent UV light irradiation. The irradiation of the free HRP solution induced a minor increase in the RFU due to a direct photooxidation of AUR. In contrast, the RFU of the HRP-FT was largely

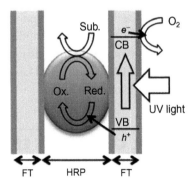

Figure 10 Schematic model of photochemical control of HRP interstratified with FT layers.

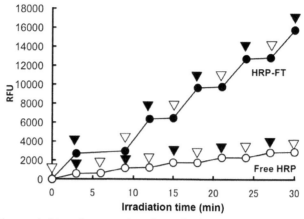

Figure 11 Photoswitching of enzymatic activity of unbound HRP (open circles) and HRP intercalated into FT layers (closed circles). The UV light was turned on and off at points indicated by open and closed triangles, respectively. The reactant solution contained 0.025 mg/mL of HRP and 0.046 mg/mL of FT only in the data of HRP-FT. The 15 μM of AUR at pH 6.5 was employed to evaluate catalytic activity of HRP.

elevated by exposed UV light, and clear photoswitching phenomenon was observed. Furthermore, the UV light did not bring about an increase in the RFU of an FT solution without HRP, indicating negligible photocatalytic oxidation activity of the FT. These results validate the proposed photoreaction mechanism in Fig. 10. The detailed protocols are described in Kamada, Nakamura, et al. (2011).

4.2 Visible-Light-Driven Enzymatic Activity of HRP Bound to Semiconductors with a Narrow Band Gap

Even though UV light irradiation activated HRP intercalated into the FT layers, activated the enzyme exposure to UV light for a long period may deactivate the enzyme. Since the FT layers cannot absorb visible light, we used semiconductor with a narrow band gap (Kamada, Moriyasu, & Soh, 2012). We have employed an n-type semiconductor α-Fe_2O_3 with a band gap of 2.2 eV for this purpose. Thin films of α-Fe_2O_3 were fabricated through a photoelectroless deposition technique on a gold substrate in a mixed solution of $FeSO_4$ and H_2PtCl_6 (20 mL) followed by annealing at 1073 K in air. The Fe^{2+} in the raw material was oxidized to Fe^{3+} and then deposited as solid FeOOH on the gold substrate. Dissolved oxygen molecules acted as oxidizing agents. The redox reactions proceeded by electron exchange through the conductive gold substrate, that is, the local cell mechanism. FeOOH is also an n-type semiconductor, and hence the photoirradiation to the FeOOH caused band gap excitation. Holes left in the VB participated in further oxidation of Fe^{2+} and film growth (Kamada, Hyodo, & Shimizu, 2010; Kamada & Moriyasu, 2011). The Pt in the film had a tetravalent state according to X-ray photoelectron spectroscopy. Since the Pt^{4+} was occupied Fe^{3+} sites in α-Fe_2O_3, a carrier density of α-Fe_2O_3 would increase by the Pt^{4+} doping, causing enhanced photoactivity.

The thin film was immersed in an HRP solution to adsorb the film surface and adsorption process reached saturation in about 2 h. Since α-Fe_2O_3 has a high pI, the HRP solution was buffered at pH 6.5 to enhance electrostatic interaction (surface charge: α-$Fe_2O_3 > 0 >$ HRP). The HRP–α-Fe_2O_3(Pt) was immersed in the AUR solution and exposed to visible light with different wavelengths. Figure 12A compares RFU changes of bare α-Fe_2O_3(Pt) and HRP–α-Fe_2O_3(Pt) under blue light illumination (473 nm, 1 mW/cm^2). The RFU of HRP–α-Fe_2O_3(Pt) was continuously increased with the irradiation time, and the reaction rate was much higher than that of the bare α-Fe_2O_3(Pt). Namely, the blue light trigged off the enzymatic reaction of HRP. Figure 12B shows influence of irradiation intensity on the visible-light-driven enzymatic oxidation of the

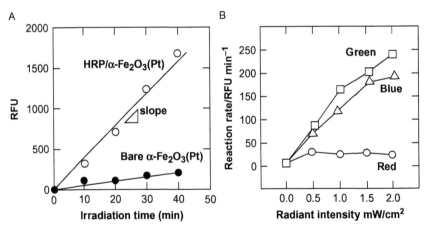

Figure 12 (A) Enhancements of relative fluorescence unit (RFU) of 0.1 mM Amplex UltraRed (pH 6) with bare or HRP-adsorbed α-Fe_2O_3(Pt) during blue light irradiation (1 mW/cm^2). (B) Effects of radiant intensity on photoinduced enzymatic reaction rate under irradiation by colored lights. *Reprinted with permission from Kamada et al., 2012. Copyright 2012 American Chemical Society.*

HRP–α-Fe$_2$O$_3$(Pt). As expected, red light (636 nm) with a photon energy smaller than the band gap of α-Fe$_2$O$_3$ did not show distinct photoenzymatic reaction. In contrast, the reaction rate under irradiation of blue or green (529 nm) light was monotonically elevated with an increase of irradiation intensity, suggesting a facile control of activity by light. Moreover, the use of visible light realized sustainable enzymatic reaction for at least 8 h. Recently, the nanometric FT layers with a narrow band gap were prepared by the hydrolysis, and then the visible-light-driven enzymatic activity has been investigated after hybridizing with HRP (Kamada, Ito, & Soh, 2015). As a result, the nanometric FT layers were also useful as host layers.

5. BIORECOGNITION USING DOPED TITANATE LAYERS MODIFIED WITH BIOMOLECULES

A number of layered oxides doped with rare earth elements show fluorescence. In addition, the layered oxides can be modified with various biomolecules that can be used for bioaffinity reaction. Hence, the modified fluorescent titanate (Eu-doped titanate) was employed to detect bioaffinity reactions (Tsukahara, Soh, & Kamada, 2012). The modification process consists of two steps as shown in Fig. 13. At first, avidin proteins were physically adsorbed on the titanate surface. Subsequently, biotinylated functional biomolecules (antibody) was trapped by avidin on the basis of a well-known strong affinity of avidin for biotin. The titanate layers modified with biotinylated antirabbit IgG antibody exhibited a practical recognition performance for rabbit IgG antigen. Since the host titanate was doped with Eu^{3+}, fluorescence detection and imaging of the recognition were achieved. Such soft surface modification of titanate layers was also useful for selective detection-specific living cells like human red blood cells when the Eu-doped titanate was coated with lectins that could be bound to specific carbohydrates on the cells.

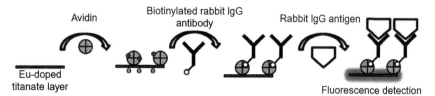

Figure 13 Biorecognition process using Eu-doped titanate layers.

6. MAGNETIC APPLICATION OF HYBRIDS COMPOSED OF ENZYMES AND DOPED TITANATES

6.1 Hybrids Composed of Enzyme and FT Layers

Generally, enzymes are expensive compared with chemical catalysts. Therefore, the development of methods for the recycling of enzymes is important. Since a separation and collection method utilizing magnets is facile, simple, and cheap, the sequential use of enzymes based on magnetic separation is one of the ideal techniques for beneficial use of enzymes. On the basis of such a background, we selected the FT layer as a host, because the FT contains a magnetic element (Fig. 14), and prepared enzyme–FT complex for the magnetic separation of the immobilized enzyme (Kamada, Tsukahara, et al., 2010).

The colloidal solution of exfoliated FT layers (Section 2.1) was mixed with an HRP solution at pH 4, because pI of HRP was estimated to be ca. 5. Intercalation of HRP into the FT layers was confirmed by XRD, absorbance, and FTIR measurements, suggesting that deformation of the secondary structure of HRP did not occur through the hybridization process. The kinetic parameters (K_m and V_{max}) for free and bound HRP (1.5 μM), in which guaiacol was selected as a substrate, were as follows: $K_m = 1.4$ mM, $V_{max} = 0.054$ mM/s for free HRP, and $K_m = 1.9$ mM, $V_{max} = 0.032$ mM/s for HRP bound to FT.

As mentioned above, the synthesized FT layers might be effective as a magnetic host material for enzyme. Several oxides doped with 3d magnetic elements show ferromagnetism even at room temperature (Osada, Ebina, Takada, & Sasaki, 2006). The magnetic character of HFT and HRP-FT

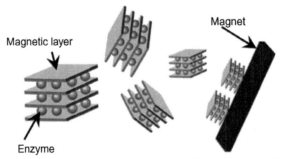

Figure 14 Magnetic separation using enzyme-intercalated magnetic layers.

at room temperature was evaluated by a vibrating sample magnetometer. The small hysteresis loops in the magnetization curves of HFT and HRP-FT indicate that the HFT and HRP-FT have magnetic properties suitable for separation. The magnetic separation of HRP-FT was performed using commercial magnets. A few ball-shaped neodymium magnets were added to the suspension of HRP-FT (0.1 mg/mL). As a result, the HRP-FT was attracted to the magnets and the suspension became colorless and transparent, indicating that the HRP-FT can be magnetically collectable. Not surprisingly, undoped layered titanate was untrapped on the magnets. On the contrary, an extremely high field generated by superconducting magnets was applied to a magnetic separation using an LDH including magnetic elements (Nakahira, Kubo, & Murase, 2007). The FT layer is a new interesting host material for magnetic separation of enzymes.

6.2 Nanohybrids Composed of Enzyme, Titanate Nanosheet, and Magnetic Beads

As mentioned in Section 6.1, FT is a novel attractive magnetic host for immobilization of enzymes. However, the magnetic property of FT is not so strong that neodymium magnets are required to be immersed into the suspension of enzyme-FT nanohybrids for magnetic separation. Therefore, we focused on magnetic beads (MBs) which are expected to show stronger magnetic property. In this study, undoped titanate was complexed with MBs and enzymes, and the new complex material has been tested for magnetic separation and sequential use of enzymes (Soh, Kaneko, Uozumi, Ueda, & Kamada, 2014).

Nanometric undoped titanate synthesized by the hydrolysis reaction, enzyme (HRP), and amino group-terminated MB (Therma-Max(R) LAm Amine (Wako Pure Chemical)) was used. First, 0.2 M titanate (50 µL), HRP (500 µg, 490 µL), and the amino group-terminated MB (4 mg/mL, 10 µL) were mixed in 20 mM acetate buffer at pH 4 (450 µL). From zeta potential measurements, the MB has positive surface charge (+16.4 mV) since the surface of the MB is covered with amino group-terminated polymer. As previously mentioned, the surface of HRP has a positive charge and that of titanate is negative at pH 4. Therefore, a complex is formed through the electrostatic interaction among the titanate, the HRP, and the MB. In fact, aggregates with light-brown color were observed when these were mixed in the buffer. After a centrifugation (10,000 rpm, 2 min), there is almost no HRP (0.3%) in the supernatant from the results of absorbance measurements. In addition, the suspension of the aggregate was attracted by commercial neodymium magnets. These results indicate that the proposed HRP/titanate/MB complex was formed under the conditions. The enzymatic activities of HRP in the HRP/titanate/MB nanohybrids were confirmed from an increase in absorption upon addition of several substrates (guaiacol, o-phenylenediamine, and pyrogallol). As stated above, the V_{max} value for the bound HRP was decreased compared with that for free HRP, probably due to the steric hindrance of the titanate/MB. On the other hand, K_m value for the bound HRP is relatively higher than that for the free HRP.

The magnetic separation of HRP/titanate/MB nanohybrids was done using commercial neodymium magnets. As a result, the HRP/titanate/MB nanohybrids were efficiently separated from the solution by external magnetic field from the commercial magnets, while HRP-FT needed the immersion of neodymium magnets to the dispersant solution for the magnetic separation. The MB has small particle size about ~100 nm and the surface is covered by thermosensitive polymer (poly(N-isopropylacrylamide)). Generally, the MB can be used for magnetic separation in the presence of salts such as NaCl at high temperature because the MBs aggregate through the phase transition of the thermosensitive polymer and enhance the magnetic property under these conditions. However, the HRP/titanate/MB could be trapped by magnets in pure water in the absence of salts at room temperature, probably because the MBs with high density in the HRP/titanate/MB complex show stronger magnetism. Absorbance measurements using o-phenylenediamine indicated that the HRP in the HRP/titanate/MB complex keeps about 80% of enzymatic activity after five times magnetic separation processes. The small decrease in enzymatic activity after the magnetic separation is probably due to loss of the HRP/titanate/MB complex during the separation processes, suggesting that there is still room for improvement in the separation method. Nevertheless, this method enabled a facile and prompt magnetic separation of enzyme and the sequential uses. On this point, the proposed method is a novel attractive technique for a beneficial use of enzymes.

7. CONCLUSIONS

This chapter summarized synthesis and several synergistic functions of nanohybrid materials composed of semiconducting layered titanate and enzyme. Enzymes were bound to the layered titanate via electrostatic interaction at an appropriate pH and then periodical microstructure of host layers and interlayer enzymes was formed. In other words, the enzymes were intercalated between the host layers. In addition to stability enhancement of enzymes at extreme conditions, enzymatic activity could be manipulated by photoirradiation that induced band gap excitation of the semiconducting titanate layers doped with Fe (FT) or α-Fe_2O_3 thin films and subsequent activation of the interlayer enzymes. The FT layer as a host for enzymes also achieved magnetic application such as magnetic separation after reaction and reuse similar to conventional immobilized enzymes. Since there are many other useful properties found in layered oxides, our future effort will be devoted to develop novel functions of the nanohybrid materials utilizing both advantages of enzymes and layered oxides.

ACKNOWLEDGMENTS
This work was partly supported by JSPS KAKENHI Grant No. 26410244 and 15K05542.

REFERENCES
Ban, T., Yoshikawa, S., & Ohya, Y. (2011). Synthesis of transparent aqueous sols of colloidal layered niobate nanocrystals at room temperature. *Journal of Colloid and Interface Science, 364*, 85–91.

Deshapriya, I. K., Kim, C. S., Novak, M. J., & Kumar, C. V. (2014). Biofunctionalization of α-zirconium phosphate nanosheets: Toward rational control of enzyme loading, affinities, activities and structure retention. *ACS Applied Materials & Interfaces, 6*, 9643–9653.

Fruk, L., Rajendran, V., Spengler, M., & Niemeyer, C. M. (2007). Light-induced triggering of peroxidase activity using quantum dots. *ChemBioChem, 8*, 2195–2198.

Harada, M., Sasaki, T., Ebina, Y., & Watanabe, M. (2002). Preparation and characterizations of Fe- or Ni-substituted titania nanosheets as photocatalysts. *Journal of Photochemistry and Photobiology A, 148*, 273–276.

Ho, K. M., Mao, X., Gu, L., & Li, P. (2008). Facile route to enzyme immobilization: Core–shell nanoenzyme particles consisting of well-defined poly(methyl methacrylate) cores and cellulase shells. *Langmuir, 24*, 11036–11042.

Ikemoto, H., Chi, Q., & Ulstrup, J. (2010). Stability and catalytic kinetics of horseradish peroxidase confined in nanoporous SBA-15. *The Journal of Physical Chemistry C, 114*, 16174–16180.

Kamada, K. (2014). Intense emission from photoproteins interacting with titanate nanosheets. *RSC Advances, 4*, 43052–43056.

Kamada, K., Hyodo, T., & Shimizu, Y. (2010). Visible-light-enhanced electroless deposition of nanostructured iron oxyhydroxide thin films. *The Journal of Physical Chemistry C, 114*, 3707–3711.

Kamada, K., Ito, D., & Soh, N. (2015). Visible-light-induced activity control of peroxidase bound to Fe-doped titanate nanosheets with nanometric lateral dimensions. *Bioconjugate Chemistry, 26*, 2161–2166.

Kamada, K., & Moriyasu, A. (2011). Photo-excited electroless deposition of semiconductor oxide thin films and their electrocatalytic properties. *Journal of Materials Chemistry, 21*, 4301–4306.

Kamada, K., Moriyasu, A., & Soh, N. (2012). Visible-light-driven enzymatic reaction of peroxidase adsorbed on doped hematite thin films. *The Journal of Physical Chemistry C, 116*, 20694–20699.

Kamada, K., Nakamura, T., & Tsukahara, S. (2011). Photoswitching of enzyme activity of horseradish peroxidase intercalated into semiconducting layers. *Chemistry of Materials, 23*, 2968–2972.

Kamada, K., & Soh, N. (2015). Enzyme-mimetic activity of Ce-intercalated titanate nanosheets. *The Journal of Physical Chemistry B, 119*, 5309–5314.

Kamada, K., Tsukahara, S., & Soh, N. (2010). Magnetically applicable layered iron-titanate intercalated with biomolecules. *Journal of Materials Chemistry, 20*, 5646–5650.

Kamada, K., Tsukahara, S., & Soh, N. (2011). Enhanced ultraviolet light tolerance of peroxidase intercalated into titanate layers. *The Journal of Physical Chemistry C, 115*, 13232–13235.

Kamada, K., Yamada, A., & Soh, N. (2015). Enhanced catalytic activity of enzymes interacting with nanometric titanate nanosheets. *Royal Society of Chemistry Advances, 5*, 85511–85516.

Kumar, C. V., & Chaudhari, A. (2000). Proteins immobilized at the galleries of layered α-zirconium phosphate: Structure and activity studies. *Journal of the American Chemical Society, 122*, 830–837.

Kumar, C. V., & Chaudhari, A. (2002). High temperature peroxidase activities of HRP and hemoglobin in the galleries of layered Zr(IV)phosphate. *Chemical Communications*, 2382–2383.

Mozhev, V. V., Khmelnitsky, Y. L., Sergeeva, M. V., Belova, A. B., Klyachko, N. L., Levashov, A. V., et al. (1989). Catalytic activity and denaturation of enzymes in water/organic cosolvent mixtures: α-Chymotrypsin and laccase in mixed water/alcohol, water/glycol and water/formamide solvents. *The FEBS Journal, 184*, 597–602.

Nakahira, A., Kubo, T., & Murase, H. (2007). Synthesis of LDH-type clay substituted with Fe and Ni ion for arsenic removal and its application to magnetic separation. *IEEE Transactions on Magnetics, 43*, 2442–2444.

Ohya, T., Nakayama, A., Ban, T., Ohya, Y., & Takahashi, Y. (2002). Synthesis and characterization of halogen-free, transparent, aqueous titanate colloidal solutions from titanium alkoxide. *Chemistry of Materials, 14*, 3082–3089.

Osada, M., Ebina, Y., Takada, K., & Sasaki, T. (2006). Gigantic magneto-optical effects in multilayer assemblies of two-dimensional titania nanosheets. *Advanced Materials, 18*, 295–299.

Soh, N., Kaneko, S., Uozumi, K., Ueda, T., & Kamada, K. (2014). Preparation of an enzyme/inorganic nanosheet/magnetic bead complex and its enzymatic activity. *Journal of Materials Science, 49*, 8010–8015.

Tsukahara, S., Soh, N., & Kamada, K. (2012). Soft surface modification of layered titanate for biorecognition. *The Journal of Physical Chemistry C, 116*, 19285–19289.

Wang, Q., Gao, Q., & Shi, J. (2004). Enhanced catalytic activity of hemoglobin in organic solvents by layered titanate immobilization. *Journal of the American Chemical Society, 126*, 14346–14347.

Zhang, Y., Chen, X., Wang, J., & Yang, W. (2008). The direct electrochemistry of glucose oxidase based on layered double-hydroxide nanosheets. *Electrochemical and Solid-State Letters, 11*, F19–F21.

Zhang, Y., Ge, J., & Liu, Z. (2015). Enhanced activity of immobilized or chemically modified enzymes. *ACS Catalysis, 5*, 4503–4513.

CHAPTER SEVEN

Bioconjugation of Antibodies and Enzyme Labels onto Magnetic Beads

B.A. Otieno*,[1], C.E. Krause[†],[1],[2], J.F. Rusling*,[‡],[§],[¶],[3]
*Department of Chemistry, University of Connecticut, Storrs, Connecticut, USA
[†]Department of Chemistry, University of Hartford, West Hartford, Connecticut, USA
[‡]Institute of Materials Science, University of Connecticut, Storrs, Connecticut, USA
[§]Department of Cell Biology, University of Connecticut Health Center, Farmington, Connecticut, USA
[¶]School of Chemistry, National University of Ireland, Galway, Ireland
[3]Corresponding author: e-mail address: james.rusling@uconn.edu

Contents

1. Introduction	136
2. Bioconjugation of Magnetic Beads	138
2.1 Preparation of Dual-Labeled Magnetic Beads	139
3. Characterization of Magnetic Bead Bioconjugates	142
3.1 2,2′-Azino-bis(3-Ethylbenzothiazoline-6-Sulfonic Acid) (ABTS) Enzymatic Assay	143
3.2 Bicinchoninic Acid Protein Assay	144
4. Integration of Magnetic Beads into Immunoassay	146
Acknowledgment	148
References	148

Abstract

Immunoassays employ antibodies and labels to capture and detect target macromolecular analytes, often from complex sample matrices such as serum, plasma, or saliva. The high affinity and specificity of antibody–antigen interactions makes immunoassays critically important analytical techniques for clinical diagnostics as well as other research applications in the areas of pharmaceutical and environmental analysis. Integration of magnetic beads (MBs) into immunoassays and other bioanalytical methodologies is a valuable approach to allow efficient target capture, enrichment, and convenient separation. In addition, large signal amplification can be achieved by preconcentration of the target and by attaching many thousands of enzyme labels to the MBs. These features have enabled MB-based biosensors to achieve ultra-low detection limits needed for advanced clinical diagnostics that are challenging or impossible using traditional immunoassays. MBs are employed either as mobile substrates for target analyte capture,

---

[1] These authors contributed equally to this chapter.
[2] Previous graduate student of University of Connecticut.

as detection labels (or label carriers), or simultaneously as substrates and labels. For optimal assay performance, it is crucial to apply an easy, efficient, and robust bead-probe conjugation protocol, and to thoroughly characterize the bioconjugated products. Herein, we describe methods used in our laboratory to functionalize MBs with antibodies and enzyme labels for ultrasensitive detection of protein analytes. We also present detailed strategies for characterizing the MB bioconjugates.

1. INTRODUCTION

Novel technologies are greatly needed to enhance laboratory productivity and provide rapid, accurate, and sensitive methods for detection of various macromolecules such as protein biomarkers, DNA, and small molecules for early disease diagnosis (Konry et al., 2012). Recent advances in nanotechnology, in particular in the areas of nanomaterials, have advanced immunoassays toward the development of new generation of point-of-care devices that could significantly enhance test speed and sensitivity (Rusling, Bishop, Doan, & Papadimitrakopoulos, 2014). Integration of nanomaterials, such as magnetic beads (MBs), quantum dots, and carbon nanotubes, into immunoassays as detection labels, or as substrates onto which target is captured have important advantages (Chikkaveeraiah, Bhirde, Morgan, Eden, & Chen, 2012; Tekin & Gijs, 2013). Nanomaterials of various sizes, shapes, and compositions have been developed and extensively utilized, providing exciting possibilities for rapid and highly sensitive detection systems (Sapsford et al., 2013).

Among these nanomaterials, MBs have attracted much attention in immunoassays owing to their unique properties. The most useful MBs for bioconjugation feature a magnetic core surrounded by a nonmagnetic polymer coating for attachment of biomolecules (Fig. 1). Iron oxides such as magnetite (Fe_3O_4) or maghemite (γFe_2O_3) are preferentially used as core materials due to their stability. The outer polymer coating serves to add surface functional coating to the beads and protects the metal oxide from external media (Llandro, Palfreyman, Ionescu, & Barnes, 2010; Mani, Chikkaveeraiah, & Rusling, 2011). To prevent agglomeration, superparamagnetic nanoparticles are usually dispersed or embedded into a matrix such as polystyrene, silica, dextran, and albumin. The most commonly employed MBs in immunoassays are superparamagnetic or non-remnant. These beads have nominally zero magnetization in the absence of a magnetic field but become magnetic in an applied magnetic field. When fabricated

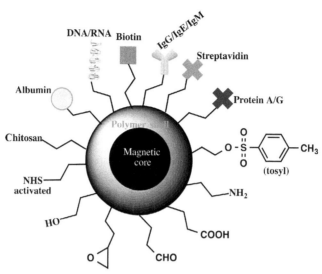

Figure 1 Variety of commercially available functionalized MBs with coatings of either organic functional groups to attach biomolecules or particular biomolecules that can bind specific moieties. (See the color plate.)

with a dense superparamagnetic nanoparticle center embedded in polymer housing, the resulting MBs behave as nonmagnetic particles in the absence of magnetic fields, although a bit of aggregation may result from small nonideal magnetic properties. A great advantage of superparamagnetic beads as opposed to nonmagnetic particles is their ease of manipulation with simple, inexpensive permanent magnets, or electromagnets. Very efficient isolation of target analytes from biomedical samples can be achieved by both manual and automated systems (Mani et al., 2011; Rusling, 2013; Rusling et al., 2014; Tekin & Gijs, 2013).

MBs of various sizes, densities, magnetic susceptibilities, material composition, and a wide variety of surface chemistries are commercially available from companies such as Invitrogen, Solulink, Micromod, Bangs Labs, and Merck with diameter ranging from 10 to 50 μm. The user can choose the desired surface chemistry to link the molecules of interest. MBs conjugated with antibodies provide large surface area per unit volume to enable efficient capture of target analytes. As a result, the sensitivity of the assays is increased due to the high efficiency of interaction between the sample and reagents that can provide large preconcentration factors. Moreover, MBs can be easily dispersed into the sample matrix with gentle shaking, hence faster binding kinetics can be achieved (Chikkaveeraiah et al., 2012; Mani et al., 2011). In addition to the ease of manipulation, shorter assay time, and high sensitivity, MBs are compatible with diverse detection and signal processing approaches, including electrochemiluminescence (Kadimisetty et al., 2015; Sardesai, Kadimisetty, Faria, & Rusling, 2013), surface plasmon resonance (Joshi, Peczuh, Kumar, & Rusling, 2014; Krishnan, Mani, Wasalathanthri, Kumar, & Rusling, 2011), and electrochemical detection (Rusling, 2013; Rusling et al., 2014; Tekin & Gijs, 2013).

Our laboratory has exploited MBs conjugated with massive copies of antibodies and enzyme labels in electrochemical immunosensors for a series of prostate and oral cancer biomarkers (Chikkaveeraiah, Mani, Patel, Gutkind, & Rusling, 2011; Krause et al., 2013; Malhotra et al., 2012; Mani, Chikkaveeraiah, Patel, Gutkind, & Rusling, 2009; Otieno et al., 2014). MBs conjugated with tens of thousands of detection antibodies (Ab_2) and up to a half million horseradish peroxidase (HRP) labels have been used to enhance the sensitivity of the immunoassays (Chikkaveeraiah et al., 2011; Krause et al., 2013, 2015; Malhotra et al., 2012; Otieno et al., 2014). Ab_2 and HRP attachment onto MBs, and purification of the bead bioconjugates (MB–Ab_2–HRP), is facilitated by magnetic separations. When integrated with microfluidic systems, position of the resulting bioconjugate beads can be magnetically controlled to facilitate faster assay times. Heavily labeled detection particles have led to detection limits in low fg/mL range, enhancing sensitivity 100–1000 times as opposed to conventional immunoassays using singly labeled Ab_2's (Jensen et al., 2009; Mani et al., 2009; Rusling, 2013). High local concentrations of antibodies on the beads greatly favor the binding of protein analytes (Mani, Wasalathanthri, Joshi, Kumar, & Rusling, 2012), while the multiple enzyme labels amplify signals (Chikkaveeraiah et al., 2012; Lei & Ju, 2012). However, in order to obtain the high sensitivity and reproducibility of immunoassays, an easy, efficient, and robust bead-probe conjugation protocol needs to be employed.

2. BIOCONJUGATION OF MAGNETIC BEADS

For optimal bioanalytical performance, it is crucial to employ a suitable and reproducible method for bioconjugation of MBs with specific recognition and signal-triggering elements. Choice of conjugation strategy is dictated by a combination of factors including MBs size, nature of the MB surface coating, available functional groups, or the type of biological molecule and its chemical composition (Lei & Ju, 2012; Sapsford et al., 2013). Commercial MBs are available with a wide variety of functional groups such as amine, carboxyl, epoxy, tosyl, hydroxyl, N-hydroxysuccinimide as well as biological molecules such as biotin, streptavidin, protein A, protein G, and antibodies (Fig. 1; Llandro et al., 2010; Mani et al., 2011).

Biomolecules can be conjugated to the surface coatings of MBs either directly, or using cross-linkers or other reagents (see Fig. 1). Bioconjugation approaches usually include noncovalent interaction such as physical adsorption, specific affinity interaction, entrapment of molecules around the MBs, and covalent interaction of biomolecules with the functional groups on MBs surface (Lei & Ju, 2012; Sapsford et al., 2013). Functional groups can be activated for coupling using EDC–NHSS chemistry for carboxylates or glutaraldehyde for amines to attach appropriate functional groups of biomolecules. Particles precoated with streptavidin can capture biotin-labeled biomolecules. Surface tosyl-, NHS-activated, and epoxide groups on MBs can be used to attach biomolecules directly without crosslinking agents. Protein A-coated particles can selectively bind to Fc regions of antibodies for oriented immobilization (Mani et al., 2011).

Herein, we provide detailed protocols used in our laboratory to functionalize MBs with antibodies as recognition elements and HRP as a signal-triggering element, also known as dual-labeling of MBs, for ultrasensitive detection of protein cancer biomarkers. We provide protocols used for conjugation of streptavidin-coated MBs as well as tosyl-activated MBs and highlight the strategies employed for optimization and characterization of the bioconjugates.

2.1 Preparation of Dual-Labeled Magnetic Beads
2.1.1 Tosyl-Activated Magnetic Beads

Tosyl-activated MBs provide reactive sulfonyl esters that covalently link antibodies or other ligands that contain primary amino or sulfhydryl groups to the magnetic particle surface (Fig. 2). Immobilization of antibodies on these MBs occurs via the Fc region, thus ensuring optimal orientation of the antibody, which increases the capture yield of the target analyte. The physical adsorption of the antibodies to MB is rapid; however, the formation of covalent bonds requires longer time. Therefore, preparation of these conjugates should take place at 37 °C for 24 h. Additional incubation with blocking buffers containing 0.01–0.5% bovine serum albumin (BSA) solution will aid in minimizing nonspecific binding (NSB) and increase the functionality of the coupled antibodies and enzyme labels. Buffers used for the preparation of antibody conjugates should be free of any reactive groups including amines, thiols, and hydroxyls as well as any sugars and stabilizers

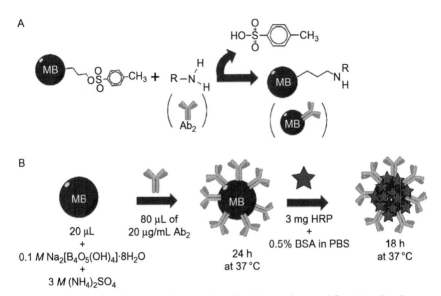

Figure 2 (A) The covalent attachment of antibodies to the tosyl functionalized magnetic particles. The tosyl groups act as leaving groups for surface amine groups present on antibodies for covalent attachment. (B) The complete conjugation protocol for both the attachment of antibodies as well as HRP enzyme labels. (See the color plate.)

that may interfere with binding. To facilitate the coupling efficiency, buffers with high ionic strength should be used as they promote hydrophobic binding. In addition, tosyl groups are more reactive at higher pH, therefore, sodium borate buffer (pH 9.5) should be used.

1. Vortex the slurry of the tosyl-activated MBs (Dynabeads® MyOne™ Tosyl-activated product no. 65501) to homogeneity before dispensing.
2. Transfer 20 μL (100 mg/mL) of tosyl-activated MBs to a 1.5 mL microcentrifuge tube and suspend in 600 μL of 0.1 M sodium borate buffer (pH 9.5).
3. Place the microcentrifuge tube on a magnet (Invitrogen Dynal magnet) for 2 min and discard the supernatant. Resuspend the tosyl-activated MBs in 600 μL of 0.1 M sodium borate buffer (pH 9.5).
4. Repeat step 3 once more to ensure the excess sulfonyl esters are removed from tosyl-activated MBs. Once supernatant is removed, resuspend MBs in 290 μL of 0.1 M sodium borate buffer (pH 9.5), 207 μL of 3 M ammonium sulfate (pH 9.5), and 80 μL (0.02–1 mg/mL) of desired concentration of antibody (Ab_2), all into the microcentrifuge tube.
5. Incubate the microcentrifuge tube with its contents (MB–Ab_2) at 37 °C for 24 h with slow tilt rotation (Invitrogen Dynabeads MX mixer).

6. Following incubation, place the microcentrifuge tube on a magnet (Invitrogen Dynal magnet) for 2 min and discard the supernatant, removing any excess unbound antibody (Ab_2). Resuspend the bioconjugates with 600 µL of 20 mM phosphate buffer containing detergent Tween-20 (pH 7.4).
7. Place the microcentrifuge tube, again, on the magnet (Invitrogen Dynal magnet) for 2 min and discard the supernatant. Resuspend the MB–Ab_2 with 625 µL of 3 mg HRP in 0.5% BSA in phosphate buffer (pH 7.4).
8. Incubate the bioconjugate for 18 h at 37 °C with slow rotation to attach HRP to MB–Ab_2 to form HRP–MB–Ab_2.
9. Place the BSA blocked HRP–MB–Ab_2 bioconjugates on a magnet (Invitrogen Dynal magnet) for 2 min and discard the supernatant, removing any excess unbound HRP. Resuspend the HRP–MB–Ab_2 bioconjugate in 625 µL of 0.1% BSA in phosphate buffer (pH 7.4).
10. Repeat wash (step 9) three times. Once sufficiently washed, resuspend the HRP–MB–Ab_2 bioconjugates in 625 µL of 0.1% BSA in phosphate buffer (pH 7.4). Place the conjugates at 4 °C until further use. These HRP–MB–Ab_2 samples can be stored for 2–3 weeks without noticeable degradation in performance (Chikkaveeraiah et al., 2011).

2.1.2 Streptavidin-Coated Magnetic Beads

Streptavidin MBs contain a monolayer of streptavidin covalently coupled to the surface of the beads. This streptavidin monolayer has a high affinity for biotinylated biomolecules (Fig. 3). The binding of streptavidin to biotin is one of the strongest known noncovalent biological interactions with femtomolar affinity constants (Howarth et al., 2006). Once the streptavidin tetramer is bound to the surface of the MBs, there are two or three

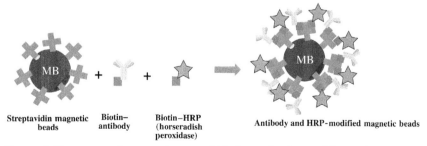

Figure 3 The noncovalent attachment of biotin–antibodies and biotin–horseradish peroxidase labels to the surface of streptavidin-coated MBs. (See the color plate.)

biotin-binding sites available for each streptavidin molecule on the surface of the bead. Streptavidin's high affinity for biotin enables biotinylated biomolecules to be captured within 25 min at 37 °C, this greatly reduces conjugate preparation times. We recommend a ratio of 1:2:4 for MBs:Ab$_2$:HRP when preparing the streptavidin MB conjugates.

1. Vortex the medium slurry of the streptavidin MBs (Dynabeads® MyOne™ streptavidin T1 product no. 65601) to homogeneity before dispensing.
2. Transfer 20 μL (10 mg/mL) of streptavidin MBs to a 1.5 mL microcentrifuge tube and suspend in 200 μL of 0.1% BSA in phosphate buffer (pH 7.4).
3. Place the microcentrifuge tube on a magnet (Invitrogen Dynal magnet) for 2 min and discard the supernatant. Resuspend the streptavidin MBs in 200 μL of 0.1% BSA in phosphate buffer (pH 7.4).
4. Repeat (step 3) two more times to ensure the excess storage solution is removed from streptavidin MBs. Once supernatant is discarded from the third wash, resuspend MBs in 80 μL of 0.1% BSA in phosphate buffer (pH 7.4), 40 μL (0.02–1 mg/mL) of desired concentration of antibody (Ab$_2$), and 80 μL of biotinylated-HRP (2.5 mg/mL), all into the microcentrifuge tube.
5. Incubate the microcentrifuge tube with its contents (streptavidin–MB–biotin–AB$_2$–biotin–HRP) at 37 °C for 25 min with slow tilt rotation (Invitrogen Dynabeads MX mixer).
6. Place HRP–MB–Ab$_2$ bioconjugates on a magnet (Invitrogen Dynal magnet) for 2 min and discard the supernatant. Resuspend the HRP–MB–Ab$_2$ bioconjugate in 200 μL of 0.1% BSA in phosphate buffer (pH 7.4).
7. Repeat (step 6) three times to remove any excess unbound Ab$_2$ and HRPs. Once sufficiently washed, resuspend the HRP–MB–Ab$_2$ bioconjugates in 200 μL of 0.1% BSA in phosphate buffer (pH 7.4). Place the conjugates at 4 °C until further use. The HRP–MB–Ab$_2$ can be used for 2 weeks without noticeable degradation in performance (Malhotra et al., 2012).

3. CHARACTERIZATION OF MAGNETIC BEAD BIOCONJUGATES

To establish control of MB bioconjugate preparation, the conjugates need to be characterized in order to determine the amount of both the antibodies as well as signal-generating enzyme labels. For measuring HRP activity and its content, we recommend the use of 2,2′-azino-bis(3-ethylbenzothiazoline-6-sulfonic acid) (ABTS) enzymatic assay. The bicinchoninic acid (BCA) protein assay can be used to determine the total number of active HRP and Ab_2 bound to MBs, respectively. Since BSA is an interfering agent in BCA assay, the last washing step in the bead preparation protocol should be performed using PBS buffer and not 0.1% BSA, prior to characterization to avoid measuring residual amounts of BSA.

3.1 2,2′-Azino-bis(3-Ethylbenzothiazoline-6-Sulfonic Acid) (ABTS) Enzymatic Assay

The ABTS assay is a colorimetric assay based on the ABTS cation radical formation (Keesey, 1987; Pütter & Becker, 1983). The radical formation is catalyzed by the reduction of HRP in the presence of hydrogen peroxide (Fig. 4). The ABTS cation radical exhibits a change of color from slightly yellow to an intensely turquoise colored solution with an absorbance at 405 nm. Employing this assay, we are able to determine the number of active signal-generating HRP labels present on the MB bioconjugates.

1. Prepare tosyl-activated and streptavidin-coated bead conjugate (HRP–MB–Ab_2) as outline above and also a set of the tosyl-activated bead conjugate without HRP (MB–Ab_2). The last washing step should be performed using 20 mM phosphate buffer.
2. Prepare 100 mM potassium phosphate buffer in deionized water at pH 5.0 at room temperature.
3. Prepare 9.1 mM ABTS (Sigma-Aldrich, product no. A9941) in potassium phosphate buffer, pH 5.0 (step 2).
4. Prepare 0.3% (w/w) hydrogen peroxide solution in deionized water.
5. Set spectrophotometer to kinetic mode at 405 nm calculating initial rate every second from 5 to 120 s.
6. Blank the spectrophotometer with potassium phosphate buffer (step 2).

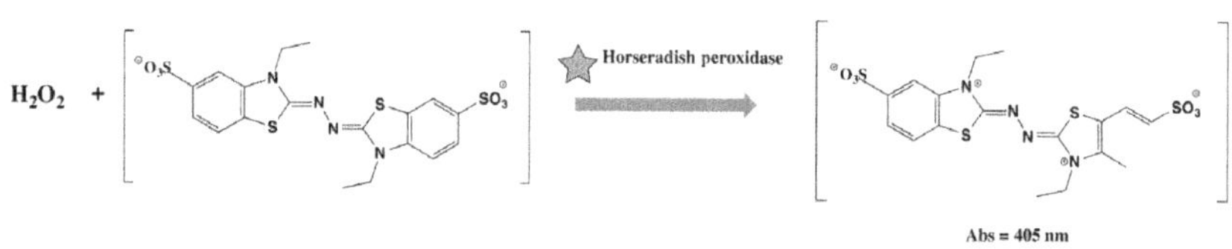

Figure 4 Formation of the 2,2′-azino-bis(3-ethylbenzothiazoline-6-sulfonic acid) (ABTS) radical catalyzed by horseradish peroxidase in the presence of hydrogen peroxide.

7. Immediately mix the reagents as indicated below by inversion and record the ΔAbsorbance for approximately 2 min.

Reagents/Sample	Test	Blank
9.1 mM ABTS solution	2.9 mL	2.9 mL
PBS buffer or MBs in PBS buffer	–	50 µL
MB–Ab$_2$ (tosyl-activated beads)	50 µL	–
HRP–MB–Ab$_2$ (tosyl and streptavidin beads)	50 µL	–
0.3% H$_2$O$_2$	100 µL	100 µL

8. The number of HRPs per MB is determined by performing the calculations below, where 3.05 is the total volume in mL of the assay, d.f. is the dilution factor (10–50 ×), 36.8 is the mM extinction coefficient of oxidized ABTS at 405 nm and 0.05 is the volume in mL of enzyme used.

$$\text{Units/mL enzyme} = \frac{[\Delta_{405/\text{min}}\text{ test}] - [\Delta_{405/\text{min}}\text{ control}] \times 3.05 \times \text{d.f.}}{(36.8)(0.05)}$$

Units/mL enzyme is then converted to the moles of HRP using pyrogallol units in HRP, Avogadro number, and molecular weight. The number of HRP per MB is obtained by diving the moles of HRP by beads per mg of the MBs (10^{12} beads/mg for tosyl-activated beads and 10^8 beads/mg of streptavidin-coated beads).

3.2 Bicinchoninic Acid Protein Assay

Bicinchoninic acid (BCA) assay or Smith assay is a copper-based colorimetric assay for total protein quantification. BCA rely on the formation of a Cu^{2+}–protein complex in a basic environment, followed by reduction of the Cu^{2+} to Cu^+ (Smith et al., 1985). The amount of Cu^{2+} that is reduced is proportional to the amount of protein present in solution. Basically, two molecules of BCA chelate to each Cu^+ ion causing a change of color from green to purple with a strong absorbance at 562 nm (Fig. 5). The bicinchoninic Cu^+ complex is influenced by both the number of peptide bonds, as well as the presence of amino acids cysteine, cystine, tyrosine, and tryptophan side chains (Wiechelman, Braun, & Fitzpatrick, 1988). Elevated temperatures increase exposure of amino acids and minimize the differences caused by unequal amino acid composition in different protein

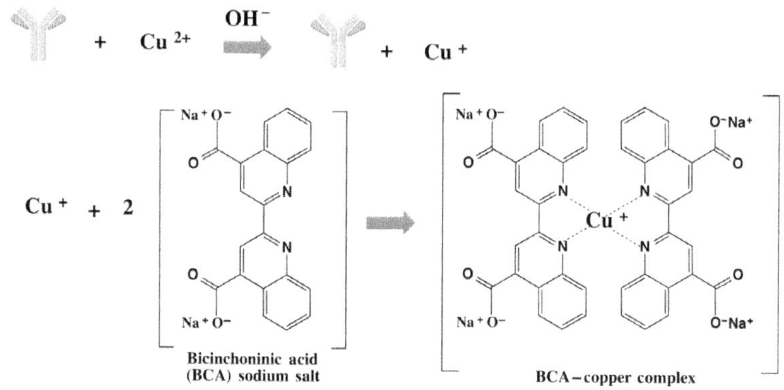

Figure 5 The formation of the bicinchoninic acid (BCA)–copper complex for the BCA total protein assay. This assay proceeds in two steps: the first being the reduction of Cu^{2+} by antibodies in a basic environment and the second step involves the reduced Cu^+ chelating with two molecules of bicinchoninic acid.

samples. Therefore, to increase assay sensitivity, the assay should be performed at an elevated temperature of 60 °C.

1. Prepare standard concentrations of detection antibody (Ab_2) with concentrations ranging from 1.25 to 20 µg/mL.
2. Formulate dilutions of MB–Ab_2 for tosyl-activated beads and HRP–MB–Ab_2 for tosyl-activated and streptavidin-coated beads (5–20 dilution factors).
3. Mix a stock solution of the BCA reagents, Pierce™ BCA Protein Assay Kit product no. 23225 (25 parts of reagent MA + 24 parts of reagent MB + 1 part reagent MC).
4. Add 500 µL of the samples in steps 1 and 2 and PBS buffer or MBs as blank to 500 µL of the stock solution of BCA in step 3.
5. Incubate the samples in step 4 at 60 °C for 1 h for the complex formation.
6. Cool all the samples to room temperature.
7. Measure the absorbance of all the samples at 562 nm within 10 min (Chikkaveeraiah et al., 2011; Smith et al., 1985; Fig. 6).
8. By subtracting absorbance of MB from that of MB–Ab_2 and HRP–MB–Ab_2, the unknown concentration of antibodies on the MB can be found from the calibration plot of antibody standards obtained with the BCA kit (Fig. 7). The number of antibodies on MP surface is obtained by dividing the number of antibodies in the dispersion by the number of particles in the dispersion (10^{12} beads/mg for tosyl-activated beads and 10^8 beads/mg of streptavidin-coated beads).

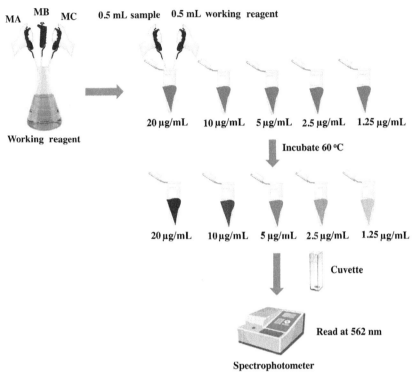

Figure 6 The workflow for the bicinchoninic acid (BCA) total protein assay including the formation of the working reagent and the development of a standard curve from a range of antibody concentrations.

Figure 7 Calculating the total antibody concentration from subtracting the absorbance of the magnetic beads from that of the magnetic beads containing the antibody.

4. INTEGRATION OF MAGNETIC BEADS INTO IMMUNOASSAY

Sensitive and selective immunoassays rely on the ability to fish out a protein analyte of interest from an ocean containing thousands of other proteins. MBs conjugated with highly selective antibodies as well as other protein capture agents provide a simple and effective way to achieve this goal. Typically, the dual-labeled MBs are added to a fluid sample, the sample is then mixed in order for proteins to be selectively captured by antibodies that are present on the MBs. Once protein analytes are captured, any interfering biomolecules present on the bead surface are removed through washing protocols involving magnetic separation that is either manual (Chikkaveeraiah et al., 2011; Krause et al., 2013; Malhotra et al., 2012) or automated (Krause et al., 2015; Otieno et al., 2014). Labels present on the bead surface such as HRP are then detected to measure the selected protein analyte of interest.

A major factor influencing the selectivity and sensitivity of immunoassays is NSB. NSB occurs when MB bioconjugates binds to nonantigen sites on the sensor during immunoassay fabrication. Generally, MB bioconjugates generate a signal even if it is bound to sites other than the analyte protein capture antibody complex. Therefore, signals arising from NSB cannot be differentiated from those originating from antigen–antibody binding and are not proportional to the analyte concentrations. As a result, NSB raises the detection limit and greatly decreases the sensitivity of the assay. Solving this problem involves creating a sensor surface that inhibits any other binding other than that of the protein antigen with its antibodies. BSA or casein with small quantity of detergent such as T20 in wash buffers are commonly employed to minimize NSB in assays. However, there is no universal blocking agent for NSB in any protein detection system; therefore, a thorough characterization of the amplifying formulation is needed to avoid NSB in any amplification strategy. Employing 2% BSA in PBS–T20 on the sensors and 0.1% BSA in PBS-T20 on MB bioconjugates have greatly inhibited NSB in our assays enabling ultra-low detection limits to be achieved (Krause et al., 2013, 2015; Malhotra et al., 2012; Otieno et al., 2014).

In order to achieve the ultra-low detection limits in immunoassays, the concentration of antibodies should be optimized on the MB bioconjugates (HRP–MB–Ab_2). Usually, MB bioconjugates are prepared with a wide range of concentrations for detection antibody (Ab_2) and tested for their immunoassay performance. Keeping the concentration of capture antibody (Ab_1) on the detection platform constant, a blank and a standard sample are tested with different MB bioconjugates with varying concentrations of Ab_2. The MB bioconjugates with the highest difference between the blank and the sample is then chosen for immunoassay applications (Krause et al., 2015). The Ab_1 on the detection platform is optimized as well by varying the concentration of Ab_1 and keeping the concentration of Ab_2 on the MB bioconjugates constant. The key factors that play a role in the MB bioconjugate with optimal concentration of antibodies are orientation and binding capacity.

Integrating these MB-based immunoassays in microfluidic systems adds to many key advantages in development of a high performance assay as these systems can reproducibly deliver these MB bioconjugates to desired locations in a simple and rapid manner. However, MBs may adsorb to microfluidic device surfaces due to these micrometer-dimensioned channels and micron tube diameters, leading to clogged channels and an increase in flow resistance (Ng, Udduyasankar, & Wheeler, 2010). In order to minimize the risk of clogging and decreased assay performance, it is recommended that microfluidic tubing be changed every few months and device systems be thoroughly washed with detergents after each use. We also recommend the use of four to five times diluted MB bioconjugate (HRP–MB–Ab$_2$) in immunoassays.

The development of MB bioconjugates for protein capture, manipulation, transport, and labeling has enabled the detection of a library of known cancer biomarker proteins. These proteins are detected at clinically relevant serum levels and have clearly demonstrated their utility in both manual and automated capture systems. There is no doubt that these MB-based technologies are/will be important tools for future protein detection systems. Interfacing these relatively simple protocols with microfluidic or other automated sample handling technologies will further propel MB protein detection systems into the clinical setting.

ACKNOWLEDGMENT

Development of protocols described herein as well as preparation of this chapter was supported by grants EB016707 and EB014586 from the National Institute of Biomedical Imaging and Bioengineering (NIBIB), NIH, USA.

REFERENCES

Chikkaveeraiah, B. V., Bhirde, A. A., Morgan, N. Y., Eden, H. S., & Chen, X. (2012). Electrochemical immunosensors for detection of cancer protein biomarkers. *ACS Nano, 6*, 6546–6561.

Chikkaveeraiah, B. V., Mani, V., Patel, V., Gutkind, J. S., & Rusling, J. F. (2011). Microfluidic electrochemical immunoarray for ultrasensitive detection of two cancer biomarker proteins in serum. *Biosensors & Bioelectronics, 26*, 4477–4483.

Howarth, M., Chinnapen, D. J.-F., Gerrow, K., Dorrestein, P. C., Grandy, M. R., & Kelleher, N. L. (2006). A monovalent streptavidin with a single femtomolar biotin binding site. *Nature Methods, 3*, 267–273.

Jensen, G. C., Yu, X., Gong, J. D., Munge, B., Bhirde, A., Kim, S. N., et al. (2009). Characterization of multienzyme-antibody-carbon nanotube bioconjugates for immunosensors. *Journal of Nanoscience and Nanotechnology, 9*, 249–255.

Joshi, A. A., Peczuh, M. W., Kumar, C. V., & Rusling, J. F. (2014). Ultrasensitive carbohydrate-peptide SPR imaging microarray for diagnosing IgE mediated peanut allergy. *Analyst, 139*, 5728–5733.

Kadimisetty, K., Malla, S., Sardesai, N. P., Joshi, A. A., Faria, R. C., Lee, N. H., et al. (2015). Automated multiplexed ECL Immunoarrays for cancer biomarker proteins. *Analytical Chemistry, 87*, 4472–4478.

Keesey, J. (1987). *Biochemical information* (1st ed.). Indianapolis, IN: Boehringer Mannheim Biochemicals, pp. 58.

Konry, T., Bale, S. S., Bhushan, A., Shen, K., Seker, E., Polyak, B., et al. (2012). Particles and microfluidics merged: Perspectives of highly sensitive diagnostic detection. *Mikrochimica Acta, 176*, 251–269.

Krause, C. E., Otieno, B. A., Bishop, G. W., Phadke, G., Choquette, L., Lalla, R. V., et al. (2015). Ultrasensitive microfluidic array for serum pro-inflammatory cytokines and C-reactive protein to assess oral mucositis risk in cancer patients. *Analytical and Bioanalytical Chemistry, 407*(23), 7239–7243.

Krause, C. E., Otieno, B. A., Latus, A., Faria, R. C., Patel, V., Gutkind, J. S., et al. (2013). Rapid microfluidic immunoassays of cancer biomarker proteins using disposable inkjet-printed gold nanoparticle arrays. *ChemistryOpen, 2*, 141–145.

Krishnan, S., Mani, V., Wasalathanthri, D., Kumar, C. V., & Rusling, J. F. (2011). Attomolar detection of a cancer biomarker protein in serum by surface plasmon resonance using superparamagnetic particle labels. *Angewandte Chemie International Edition in English, 50*, 1175–1178.

Lei, J., & Ju, H. (2012). Signal amplification using functional nanomaterials for biosensing. *Chemical Society Reviews, 41*, 2122–2134.

Llandro, J., Palfreyman, J. J., Ionescu, A., & Barnes, C. H. W. (2010). Magnetic biosensor technologies for medical applications: A review. *Medical & Biological Engineering & Computing, 48*, 977–998.

Malhotra, R., Patel, V., Chikkaveeraiah, B. V., Munge, B. S., Cheong, S. C., Zain, R. B., et al. (2012). Ultrasensitive detection of cancer biomarkers in the clinic by use of a nanostructured microfluidic array. *Analytical Chemistry, 84*, 6249–6255.

Mani, V., Chikkaveeraiah, B. V., Patel, V., Gutkind, J. S., & Rusling, J. F. (2009). Ultrasensitive immunosensor for cancer biomarker proteins using gold nanoparticle film electrodes and multienzyme-particle amplification. *ACS Nano, 3*, 585–594.

Mani, V., Chikkaveeraiah, B. V., & Rusling, J. F. (2011). Magnetic particles in ultrasensitive biomarker protein measurements for cancer detection and monitoring. *Expert Opinion on Medical Diagnostics, 5*, 381–391.

Mani, V., Wasalathanthri, D. P., Joshi, A. A., Kumar, C. V., & Rusling, J. F. (2012). Highly efficient binding of paramagnetic beads bioconjugated with 100,000 or more antibodies to protein-coated surfaces. *Analytical Chemistry, 84*, 10485–10491.

Ng, A. H. C., Udduyasankar, U., & Wheeler, A. R. (2010). Immunoassays in microfluidic systems. *Analytical and Bioanalytical Chemistry, 397*, 991–1007.

Otieno, B. A., Krause, C. E., Latus, A., Chikkaveeraiah, B. V., Faria, R. C., & Rusling, J. F. (2014). On-line protein capture on magnetic beads for ultrasensitive microfluidic immunoassays of cancer biomarkers. *Biosensors & Bioelectronics, 53*, 268–274.

Pütter, J., & Becker, R. (1983). In H. U. Bergmeyer (Ed.), *Methods of enzymatic analysis: Vol. 3* (3rd ed., pp. 286–293). Deerfield Beach, FL: Verlag Chemie.

Rusling, J. F. (2013). Multiplexed electrochemical protein detection and translation to personalized cancer diagnostics. *Analytical Chemistry, 85*, 5304–5310.

Rusling, J. F., Bishop, G. W., Doan, N., & Papadimitrakopoulos, F. (2014). Nanomaterials and biomaterials in electrochemical arrays for protein detection. *Journal of Materials Chemistry B, 2*, 12–30.

Sapsford, K. E., Algar, W. R., Berti, L., Gemmill, K. B., Casey, B. J., Oh, E., et al. (2013). Functionalizing nanoparticles with biological molecules: Developing chemistries that facilitate nanotechnology. *Chemical Reviews, 113*, 1904–2074.

Sardesai, N. P., Kadimisetty, K., Faria, R., & Rusling, J. F. (2013). A microfluidic electrochemiluminescent device for detecting cancer biomarker proteins. *Analytical and Bioanalytical Chemistry, 405*, 3831–3838.

Smith, P. K., Krohn, R. I., Hermanson, G. T., Mallia, A. K., Gartner, F. H., Provenzano, M. D., et al. (1985). Measurement of protein using bicinchoninic acid. *Analytical Biochemistry, 150*, 76–85.

Tekin, H. C., & Gijs, M. A. (2013). Ultrasensitive protein detection: A case for microfluidic magnetic bead-based assays. *Lab on a Chip, 13*, 4711–4739.

Wiechelman, K., Braun, R., & Fitzpatrick, J. (1988). Investigation of the bicinchoninic acid protein assay: Identification of the groups responsible for color formation. *Analytical Biochemistry, 175*, 231–237.

CHAPTER EIGHT

Rationally Designed, "Stable-on-the-Table" NanoBiocatalysts Bound to Zr(IV) Phosphate Nanosheets

I.K. Deshapriya[*], C.V. Kumar[*,†,‡,§,1]

[*]Department of Chemistry, University of Connecticut, Storrs, Connecticut, USA
[†]Department of Molecular and Cell Biology, University of Connecticut, Storrs, Connecticut, USA
[‡]Institute of Material Science, University of Connecticut, Storrs, Connecticut, USA
[§]Department of Inorganic and Physical Chemistry, Indian Institute of Science, Bengaluru, Karnataka, India
[1]Corresponding author: e-mail address: challa.kumar@uconn.edu

Contents

1. Introduction	152
1.1 Layered Nanomaterials	153
1.2 Overview of Our General Strategy to Stabilize Enzymes	154
1.3 Thermodynamics of Enzyme Denaturation	156
1.4 The Entropy Hypothesis	156
1.5 Our Approach to Stabilize Enzymes	157
1.6 Structure of α-ZrP	158
1.7 Advantages of Using α-ZrP for Enzyme Loading	159
2. Methods	160
2.1 Synthesis of α-ZrP	160
2.2 Exfoliation of α-ZrP Nanosheets	162
2.3 Binding of Positively Charged Enzymes to Exfoliated α-ZrP	164
2.4 Binding of Negatively Charged Proteins to Anionic α-ZrP Nanosheets	169
References	174

Abstract

Rational approaches for the control of nano–bio interfaces for enzyme stabilization are vital for engineering advanced, functional nanobiocatalysts, biosensors, implants, or "smart" drug delivery systems. This chapter presents an overview of our recent efforts on structural, functional, and mechanistic details of enzyme nanomaterials design, and describes how progress is being made by hypothesis-driven rational approaches. Interactions of a number of enzymes having wide ranges of surface charges, sizes, and functional groups with α-Zr(IV)phosphate (α-ZrP) nanosheets are carefully controlled to achieve high enzyme binding affinities, excellent loadings, significant retention of the bound enzyme structure, and high enzymatic activities. In specific cases, catalytic activities and selectivities of the nanobiocatalysts are improved over those of the corresponding pristine enzymes. Maximal enzyme structure retention has been obtained by coating the nanosheets with appropriate proteinaceous materials to soften the enzyme–nanosheet interface. These systematic manipulations are of significant importance to understand the complex behavior of enzymes at inorganic surfaces.

1. INTRODUCTION

Industrial biocatalysis is a growing area with a great promise for fine chemical synthesis, biosensing, and pharmaceutical applications. As a result, multiple approaches are flourishing to improve the stabilities of enzymes in nonbiological environments or even harsh reaction conditions (DiCosimo, McAuliffe, Poulose, & Bohlmann, 2013; Kumar & Chaudhari, 2002). Wide use of enzymes as biocatalysts in industry and laboratory is hindered by the instability of enzymes to nonbiological conditions of high temperature, wide pH ranges, or organic solvents (Schoemaker, Mink, & Wubbolts, 2003). Although advances in protein engineering allow for large-scale production of recombinant enzymes for commercial purposes (Cole-Hamilton, 2003), stabilities of these enzymes in their free form need significant improvements, and recombinant enzymes are still very expensive. Enzymes bound to nanomaterials, on the other hand, can be stabilized against nonbiological conditions and readily recycled to reduce cost (Betancor & Luckarift, 2008; Costantino et al., 2000; Mudhivarthi, Bhambhani, & Kumar, 2007; Pattammattel, Puglia, et al., 2013; Puri, Barrow, & Verma, 2013; Schmid et al., 2001). Enzyme–nanoparticle covalent conjugates, for example, show improved storage and operational stability toward heat, organic solvents, or autolysis. In some cases, they showed high storage stability even at elevated temperatures (Mateo, Palomo, Fernandez-Lorente, Guisan, & Fernandez-Lafuente, 2007). Heterogeneous nature of enzyme–nanoparticle conjugates is used to separate the biocatalyst from the reaction mixture for easy work-up. They can be recovered at the end of the reaction and then recycled to reduce the cost of the biocatalyst. In certain cases, solid-bound enzymes showed enhanced substrate selectivity, expanding the activities of ordinary enzymes toward multiple but similar substrates (Kumar & Chaudhari, 2002). Even though sufficient progress has been made in this direction, as evidenced from this volume, rational design of nanomaterials to stabilize enzymes under nonbiological conditions without compromising their activities is still an unmet current challenge.

1.1 Layered Nanomaterials

Among the many types of nanomaterials known, layered materials have been the focus of our laboratory for a variety of reasons. Layered solids provide a high surface area per unit mass of the support matrix such that a large amount of the enzyme can be deposited per unit mass (high loadings). High loadings of the biocatalyst per unit mass of the support reduce the size of the bioreactor, thereby lowering the overhead cost of the process. The high aspect ratio of layered solid (width to thickness) allows for greater access of the surface-bound enzymes to the substrates from the solution phase, facilitating the diffusional transport to the active site of the enzyme. In comparison, enzymes encased in three-dimensional solids such as silica or crystals may suffer from slow diffusion of the substrate or the product through the narrow pores of the solid. Improved diffusion with an enzyme/layered solid physical complex enhances the rate of the reaction. The surface groups of the layered solid can be functionalized to control the interactions of the enzyme with the solid, an important chemical handle to control these interactions. Enzymes trapped in the layered solid may also be protected against bacterial attack and degradation, as the interlayer spacing between the layers can be quite narrow. Thus, a number of good attributes of inorganic layered solids formed a strong basis for the rational choice for their selection as the support matrix for a variety of enzymes in our studies.

Specifically, our research group has been addressing the challenges of stabilizing enzymes with inorganic solids, while preserving their enzymatic activities, by using the nanosheets of α-Zr(IV) phosphate (α-ZrP) (Scheme 1). These are two-dimensional inorganic materials with hydrophilic surfaces, which are rich in hydroxyl groups on their surface as well as abundant quantities of water on the surface. Hydrophilic, water-rich surfaces are expected to be more bio-friendly for enzyme binding. α-ZrP has been successfully used in our laboratory to intercalate several enzymes to produce enzyme/α-ZrP complexes, for the first time. The intercalated enzymes retained their native-like structure, normal activity, and improved stability when compared to those of the corresponding free enzymes. In order to achieve excellent enzyme loading and high stability, the bio–nano interface has been systematically examined, tested, and modified to fit the needs of specific enzymes. In this chapter, we describe the detailed protocols used in the preparation of enzyme/α-ZrP biocatalysts and their full characterization. These biocatalysts are referred to as "Stable-on-the-Table" because these can be stored at room temperature for extended periods of time without significant loss of their activities.

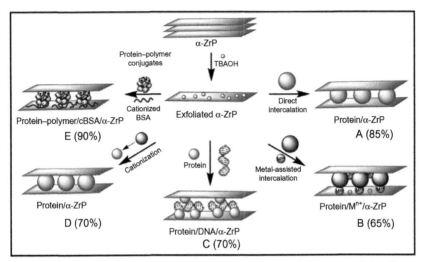

Scheme 1 Approaches (A)–(E) for controlling enzyme intercalation into α-ZrP galleries. Exfoliation of the nanosheets with tetra(n-butylammonium) hydroxide (TBA$^+$OH$^-$) followed by: (A) direct binding of positively charged enzymes to the anionic nanosheets, (B) metal ion-mediated binding of anionic enzymes to the anionic nanosheets, (C) Hb-mediated binding of anionic DNA/RNA, (D) cationization of anionic enzymes to promote their binding to the anionic nanosheets, and (E) design of protein-glues for the binding of anionic enzyme–polymer conjugates to the anionic nanosheets. A comparison of the extent of retention of hemoglobin peroxidase-like activity is given in the brackets for each of these approaches (A)–(E). (See the color plate.)

1.2 Overview of Our General Strategy to Stabilize Enzymes

There are multiple rational pathways utilized by our group to stabilize enzymes, which are summarized here. Enzymes are bound directly to the exfoliated α-ZrP nanosheets in approach A (Scheme 1), and enzyme binding triggers spontaneous assembly of the nanosheets into stacks. The resulting enzyme–nanosheet porous intercalates are protected against degradation and destabilization while allowing for access to the enzyme active site. In approach B, metal ions are used to favorably alter the net charge on negatively charged enzymes (isoelectric point, pI <7) to a net positive value, such that they are now more favorable to bind to the anionic nanosheets of α-ZrP. Under approach C, the enzyme/DNA complexes with a net positive charge or near-neutral charge are produced in the first step and then they are intercalated for enhanced stability. Depending on the amount of DNA used, the activities of the biocatalysts ranged from 70% to 110%. The charge on the α-ZrP nanosheets is strongly negative, and hence, the binding of negatively charged enzymes to these nanosheets is not strong.

Therefore, we chemically modified enzyme surface functions with polyamines to reverse enzyme charge to net positive value or cationized the enzymes, in approach D. Cationized enzymes were then readily intercalated into the nanosheet galleries. In approach E, we have formed the enzyme–poly(acrylic) acid conjugates first and then used cationized bovine serum albumin to bind these anionic enzyme–polymer conjugates to the anionic α-ZrP nanosheets. Thus, a number of different approaches are demonstrated to bind a few different enzymes or their derivatives in the galleries of inorganic nanosheets and the resulting biocatalysts indicated favorable results.

The above studies were driven by a handful of quantitative mechanistic studies of enzyme binding to nanomaterials carried out in our laboratory over a decade or more (Duff & Kumar, 2009a, 2009b). One example is improving the thermal stabilities of enzymes by intercalating them between the nanosheets of inorganic materials. A brief discussion of the enzyme denaturation thermodynamics is essential to understand the factors that control enzyme stability and this is illustrated in Scheme 2.

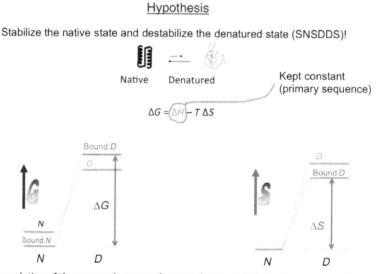

Scheme 2 Schematic diagram showing the "Entropy" hypothesis. A general approach to stabilize enzymes by lowering the conformational entropy of the denatured state. The free energy gap between the native and denatured states may be increased, which results in improved stability of the enzyme. (See the color plate.)

1.3 Thermodynamics of Enzyme Denaturation

In thermodynamic terms, the intrinsic stability of an enzyme is governed by the difference in the free energies of the native and the denatured states (ΔG_d). This is positive because the native states of most enzymes are more stable than their corresponding denatured states, at room temperature. Note that the ΔS_d term is positive because the entropy of the denatured state is greater (more disordered) than the native, ordered, state. The ΔH term is positive due to the favorable packing interactions between the individual residues that makeup the enzyme. Thus, both the ΔH and ΔS terms are positive for enzyme denaturation reaction, and both depend on T.

The ΔG_d decreases with increasing temperature, which is mostly driven by the major contributions of the $-T\Delta S_d$ term in the fundamental equation, $\Delta G_d = \Delta H_d - \Delta T S_d$. The ΔH dependence on T is very weak, while $-T\Delta S_d$ term contributes strongly to the temperature dependence of ΔG. Since denaturation temperature is when the free energies of the native and denatured states are equal (equilibrium), ΔG_d is zero at the denaturation temperature (T_d). At all temperatures below T_d, $\Delta G_d > 0$ and enzyme denaturation is not spontaneous, but at temperatures above T_d, the $\Delta G_d < 0$ and enzyme denaturation is spontaneous (Scheme 2). Therefore, the propensity to denature the enzyme depends on the sign and magnitude of ΔG_d and increase in ΔG_d can increase enzyme stability. Thus, approaches that can increase ΔG_d should increase enzyme stability.

1.4 The Entropy Hypothesis

We hypothesize that ΔG_d can be increased by stabilizing the native state of the enzyme or by destabilizing the denatured state (Scheme 2). As ΔH_d is positive and mostly depends on the primary sequence of the enzyme, one method to increase ΔG_d of ordinary enzymes is to decrease the ΔS_d. One way to decrease ΔS_d is by constraining the conformations or the number of microstates of the denatured state of the enzyme (Scheme 2). That is, if the entropy of the denatured state is lowered by some means, while keeping that of the native state unaltered, then ΔS_d can be reduced, and consequently ΔG_d can be increased. We assume that ΔH term does not change, under these conditions. Thus, a decrease in ΔS term will result in an increase in the positive value of ΔG_d, which will reduce the propensity for the enzyme to denature, and hence, increase enzyme stability.

1.5 Our Approach to Stabilize Enzymes

The above hypothesis was tested by entrapping the enzyme in the galleries of inorganic nanosheets, and the denatured state of the intercalated enzyme confined to the pseudo-two-dimensional space between the nanosheets. Hence, the conformational entropy of the denatured state of the enzyme could be lowered. Since the native state is well folded and compact, its entropy in the intercalated form will be nearly unaltered. Therefore, enzyme intercalation in the nanosheets is expected to reduce the ΔS_d and, hence, increase ΔG_d. Consequently, intercalated enzyme is expected to be more stable. This hypothesis was tested by intercalating enzymes into inorganic layered solid α-ZrP, as a model system, and by characterizing the resulting biocatalysts by a variety of physical, chemical, and biological methods.

Direct intercalation of the macromolecules such as enzymes and proteins into the galleries of the nanosheets of α-ZrP, in general, is slow due to the very narrow spacings of the interlayer region of α-ZrP, which is less than 1 nm (Garcia, Naffin, Deng, & Mallouk, 1995). However, if the nanosheets are separated or exfoliated then their surfaces will be exposed to dissolved enzymes for binding onto the nanosheet surfaces. However, such exfoliation requires sufficient free energy to move apart the layers.

In a separate set of studies, the facile exfoliation of stacks of α-ZrP nanosheets has been reported by reaction with tetra(n-butylammonium) hydroxide (TBA$^+$OH$^-$) (Kim, Keller, Mallouk, Schmitt, & Decher, 1997). Intercalation into α-ZrP is known to occur by the entry of TBA$^+$OH$^-$ molecules at the edge of the crystalline solid followed by inward diffusion along the interlamellar galleries (Kaschak et al., 1998). Insertion of the intercalant molecules in the galleries results in charge reduction, and eventual separation of weekly attractive sheets to form discrete lamellae. We adapted this approach for the exfoliation of the nanosheets and developed a general method to bind enzymes to the exfoliated α-ZrP nanosheets, for the first time (Kumar & McLendon, 1997). This major breakthrough provided new opportunities to prepare enzyme/inorganic hybrid nanomaterials, under mild conditions of pH, biologically compatible buffer, room temperature, and ionic strength. These studies facilitated experiments to test the above entropy hypothesis (Fig. 1).

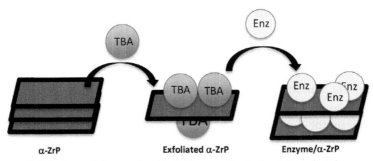

Figure 1 Exfoliation of the stacks of α-ZrP nanosheets by tetra(n-butylammonium) hydroxide (TBA⁺OH⁻) followed by enzyme binding to exfoliated α-ZrP nanosheets. Changes in interlayer spacing (d) due to intercalation were followed by powder XRD. Exfoliation causes random dispersion of layers, exposed to enzymes in the solution, and enzyme-bound nanosheets are restacked due to favorable interactions.

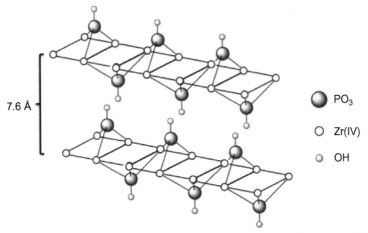

Figure 2 Schematic representation of two adjacent α-ZrP layers featuring Zr(IV) ions (*white circles*) sandwiched between two layers of phosphate anions (*blue* (*dark gray* in the print version) *spheres*). The *pink* (*light gray* in the print version) *spheres* represent ionizable OH groups on the phosphates.

1.6 Structure of α-ZrP

To understand the molecular basis for the binding of proteins to the nanosheets, a short discussion of the structure of α-ZrP and its charge characteristics is essential. The nanosheets of α-ZrP (Clearfield & Smith, 1969) are made of a plane of zirconium(IV) cations sandwiched between two layers of tetrahedral phosphate anions (Fig. 2). Three oxygens of each phosphate group are coordinated to three zirconium ions, while the remaining oxygen bears a proton, which can be replaced by various other cations, by ion exchange. Therefore, the layered structure is viewed as a 2-D matrix of phosphate anions, similar in nature to the environment supplied by phosphate buffers for biological experiments. It is also dubbed as "poor man's DNA" due to the 2-D lattice of phosphates on the nanosheets, when compared to the 1-D lattice of the DNA-phosphates.

The phosphate groups of these nanosheets, under neutral or alkaline pH conditions, are ionized and the nanosheet surface is strongly negatively charged, with a maximum of one charge per 25 Å^2 (6.6×10^{-6} mol/m^2). Consequently, cationic and positively charged enzymes are attracted to these surfaces, and the interlayer distance between the nanosheets depends on the arrangement, volume, and solvation of these counter ions (cations). When these counter ions are protons as in pristine α-ZrP, the interlayer distance is 7.6 Å.

Negatively charged, phosphate-rich, α-ZrP nanosheets facilitate strong noncovalent interactions with positively charged proteins via electrostatic and hydrogen bonding interactions (Chaudhari & Kumar, 2005). High thermal stability and acid resistance of this solid support facilitated the possibility of α-ZrP as a protecting platform for several enzymes (Lu, Wilkie, Ding, & Song, 2011). The high surface area of 100 m^2/g (Kim et al., 1997) of the solid and its charge density contribute to high enzyme loadings. The α-ZrP nanoplates pile into columns under acidic pH conditions, due to van der Waals forces between protonated phosphate anions, and these stacks of nanosheets are exfoliated prior to enzyme binding, as described earlier.

1.7 Advantages of Using α-ZrP for Enzyme Loading

There are several advantages of using layered materials for enzyme binding, α-ZrP in particular. The gallery spacings of enzyme-intercalated α-ZrP are equal to the diameter of the largest intercalated molecule, but they are small enough to shield the biomolecule from microbial attack or protease attack. Since these widths are on the order of intercalated enzyme diameters, they are wide enough to permit the diffusion of substrates and products to and from the active sites of the intercalated enzymes present in the galleries. A small fraction of enzymes may also be bound at the edges of the nanosheets and they will be much more accessible to the bulk solution than the intercalated enzymes. Unstacked nanosheets with bound enzymes provide very easy access to the substrates as well.

The layered structure of the solid also provides a convenient way to examine the enzyme/α-ZrP intercalation and binding process by powder X-ray diffraction (XRD) studies, where the expanded spacings are accurately measured. Indeed, it provided a simple and efficient method to determine the diameters of many proteins and enzymes (Kumar & Chaudhari, 2000). The charge on the nanosheets and its alteration due to enzyme binding provides a convenient way to monitor the enzyme binding to the nanosheets via zeta potential titrations (Pattammattel, Deshapriya, Chowdhury, & Kumar, 2013).

Due to the above favorable characteristics, we have used α-ZrP as a general platform for the binding of a number of different enzymes and nucleic acids with a variety of physicochemical properties (Table 1). Intercalated enzymes in α-ZrP indicated a high degree of structure and activity retention, as described below.

Here in, we report in-depth protocols used in our laboratory to synthesize and characterize α-ZrP nanosheets, their exfoliation for enzyme binding, and the formation of particular enzyme/α-ZrP complexes for biocatalytic applications. We also provide protocols for tuning the surface charge of the pristine α-ZrP with metal ions and cationized proteins to bind strongly negatively charged enzymes, which may otherwise have limited affinity for the negatively charged α-ZrP nanosheets. Thus, both cationic and anionic enzymes can be bound to these inorganic nanosheets for enhanced stability and improved biocatalytic performance.

2. METHODS

2.1 Synthesis of α-ZrP

α-ZrP synthesis was performed by reacting an aqueous solution of $ZrOCl_2$ with concentrated phosphoric acid, under refluxing conditions. Crystallinity of α-ZrP nanosheets increases with increasing concentration of phosphoric acid or increasing reflux time. The product (α-ZrP) from this reaction was isolated by precipitation as a white crystalline solid, which is thoroughly washed with water and dried to produce white powder with the 40–50% overall yield, which is suitable for enzyme binding (Kumar & Chaudhari, 2000):

$$ZrOCl_2(aq) + PO(OH)_3(aq) \rightarrow \alpha\text{-}Zr(HPO_4)_2 \cdot nH_2O(S) + HCl(aq)$$

1. Heat a sample of 12 M H_3PO_4 (concentrated acid, 50 mL) in a Pyrex round bottom (100 mL) flask to 80 °C.

    Tip: Be careful, the concentrated acid is highly corrosive and toxic.
2. Add the sample of $ZrOCl_2 \cdot 8H_2O$ (10.0 g, in 5 mL of deionized (DI) water) dropwise into the above solution while stirring and commence refluxing the mixture at 80 °C.

Table 1 Key Features of the Enzymes and Proteins Bound to the Inorganic Nanosheets Described Here

Enzyme/Property	p/	Enzyme Charge (pH 7)	Hydrodynamic Radius (Å)	Area of Cross Section (Å^2)	Molecular weight (kDa)	Number of Residues
Lysozyme	10.9	7	19.8	1232	14.3	129
Cytochrome c	10.1	7	17.2	929	12.4	104
Chymotrypsin	8.7	−1	20	1256	25	241
Myoglobin	6.9	−4	20.4	1307	16.5	153
Hemoglobin	6.6	−8	32	3217	64.5	427
Horseradish peroxidase	5.1	0	28	2463	44	308
Glucose oxidase	4.6	−62	40	5026	155	581
Bovine serum albumin	4.8	−18	33.7	3568	66.5	576
α-Lactalbumin	4.5	−7	18.8	1110	14.2	123
Tyrosinase	4.3	−8	49.1	7574	128	1126

Tip: Be sure to attach a condenser with cold water circulation. The reaction needs to be conducted in a fume hood with appropriate facilities.

3. Reflux the mixture at 80 °C for 24 h and filter the resulting white crystalline solid through a sintered glass funnel (C pore size) connected to vacuum.

    Tip: Be careful not to come in contact with the corrosive acid and other contents of the reaction mixture while filtering.

4. Wash the subsequent solid with DI water (50 mL) for several times till the wash is neutral, followed by acetone (20 mL) to remove any excess H_3PO_4 acid.

    Tip: To ensure the complete removal of acid impurities, verify the pH of the wash.

5. Finally wash the sample with generous amounts of acetone to remove residual moisture from the product and place the dry sample in an oven at 60 °C for overnight.

    Tip: Make sure that there is no acetone vapors in the sample, as this can catch fire when placed in a hot oven. Air dry the sample before placing it in the hot oven.

2.1.1 Characterization of α-ZrP by Powder XRD and FTIR

α-ZrP synthesized from the above method should be characterized by powder XRD and FTIR spectroscopy before using for enzyme binding. Crystallinity of the α-ZrP sample is confirmed by the sharp peak patterns of the powder XRD patterns which corresponds to an interlayer spacing (d-spacing) of 7.6 Å (Clearfield & Smith, 1969). See Section 2.2.1 for experimental details of measuring the d-spacings.

2.2 Exfoliation of α-ZrP Nanosheets

Layered α-ZrP, synthesized by the above method, has a layer spacing of 7.6 Å, which is overly narrow for the direct intercalation of enzymes into the galleries. The enzyme size may vary from 10 to 100 nm or even higher, and hence, the direct intercalation is very slow and may result in enzyme denaturation. We found that exfoliation of α-ZrP stacks facilitates rapid binding of enzymes to the inorganic nanosheets as these surfaces are well exposed to the solvent, under biologically compatible conditions of neutral pH and buffer conditions. Exfoliation of the nanosheets is achieved by the treatment of α-ZrP with TBA^+OH^- in which TBA^+ cations occupy a large surface area on the nanosheets. Since they are mono cationic, they create excess negative charge on the nanosheets, which might be one of the primary forces pushing the nanosheets apart (Fig. 1). This results in TBA^+-coated α-ZrP nanosheets, which are fully delaminated, and form colloidal exfoliated α-ZrP suspensions that are stable for weeks. As a result of exfoliation, the pH of the exfoliated α-ZrP increases, due to a complicated mechanism of protonation/binding. But from a practical point of view, one must adjust to pH 7 before introducing sensitive biological molecules into the medium.

1. Suspend α-ZrP (0.1 g, from Section 2.1.1) in 5 mL of DI water by sonicating the mixture for 5 min.
2. Add TBA^+OH^- (180 μL, 40%, w/w in DI water) into the solution from step 1, while stirring followed by sonication for 20 min.
3. Carefully adjust the pH of the solution from step 2 to pH 7.0 by dropwise addition of 0.1 N HCl with gentle stirring. The exfoliation process is visible, as the turbid suspension becomes translucent upon exfoliation.
4. Sonicate the translucent suspension containing 60 mM exfoliated α-ZrP from step 3 for another 10 min and dilute it with buffer as needed for enzyme binding studies. A variety of buffers can be used but make sure that the nanosheets do not precipitate in the presence of the buffer; otherwise, the enzyme binding may not occur.

    Tip: Temperature of the water bath in the sonicator should be adjusted to 25 °C to avoid heating the sample when sonicated for long times.

    Tip: Exfoliated α-ZrP can be stored in the fridge at 4 °C for a week or more, and if a solid settles at the bottom, stir and sonicate gently for short periods of time before using for enzyme binding.

2.2.1 Characterization of Exfoliated α-ZrP

Exfoliated α-ZrP synthesized from the above method is characterized by powder XRD and FTIR spectroscopy to confirm that exfoliation is complete or else there could be crystallites that will not intercalate the enzymes. As a result of exfoliation with TBA, the d-spacing increases from 7.6 to 17 Å, which is the sum of the single nanosheet thickness and a single layer of TBA^+ with an estimated average size of 9 Å.

2.2.2 Powder XRD Studies

1. Drop cast enzyme/α-ZrP suspensions (300 μL) on a clean glass slide. One might coat the glass slip with powdered α-ZrP, in the absence of the enzyme.

2. Air dry the sample covered with an inverted funnel, which controls the evaporation of the solvent and allows some time for the nanosheets to stack properly. Proper alignment of the sheets is essential to obtain a good XRD pattern with little background.
3. Scan the sample at a continuous scan rate of 1° (Rigaku Ultima IV diffractometer (Woodlands, TX)) with Cu Kα radiation ($\lambda = 0.154$ nm) beam voltage and the beam current of 40 kV and 44 mA, respectively.
4. Follow the same protocol for α-ZrP by placing powdered sample on a glass slide instead of an enzyme nanosheet suspension.

   Tip: Obtain radiation training before operating the powder XRD diffractometer. Follow all safety steps.

2.2.3 Zeta Potential Studies

The high residual charge of the α-ZrP (~ -40 mV at pH 7.0) nanosheets provides a convenient handle for monitoring enzyme binding by zeta potential studies (Deshapriya, Kim, Novak, & Kumar, 2014).
1. Make a suspension of exfoliated α-ZrP (3 mM) in DI water (pH adjusted to 7.0, 1.5 mL) in a polystyrene cuvette (Fisher Scientific).
2. Obtain the zeta potential values in triplicates at 25 °C (Brookhaven ZetaPlus zeta potential analyzer). Follow manufacturer instructions to obtain high-quality data and analysis.

2.3 Binding of Positively Charged Enzymes to Exfoliated α-ZrP

The negative charge field of exfoliated α-ZrP favors the binding of positively charged enzymes due to the favorable electrostatic attractions between protein and the anionic nanoplates. Thus, the binding experiments highly depend on the isoelectric point of the enzyme, the buffer, and the pH of the medium to be used. In our experience, both neutral and slightly negatively charged proteins also bind to the anionic surface of the nanosheets due to a complex array of interactions that control the enzyme binding (Duff & Kumar, 2009a, 2009b).
1. Prepare stock solutions of the respective enzyme (~ 5 mg/mL) in 10 mM Na$_2$HPO$_4$ buffer at pH 7 and measure the molar concentrations using absorption spectroscopy or by weight.
2. Prepare a series of enzyme/α-ZrP mixtures with increasing concentrations of the enzyme by adding exfoliated α-ZrP (6 mM) into enzymes in 2 mL Eppendorf tubes.
3. Volume up with required amounts of 10 mM Na$_2$HPO$_4$ buffer at pH 7 into each tube in order to maintain the total volume of each sample to a constant value of 2 mL.

4. Mix each sample by pipetting gently and equilibrate for 1 h at room temperature.
5. Centrifuge (Fisher Scientific Microfuge) at 12,000 rpm for 6 min and determine the unbound protein concentration in the supernatant by taking the absorbance at respective wavelengths with known extinction coefficients.
6. Estimate the amount of protein bound to α-ZrP by subtracting the free protein content from step 5 from the total protein concentration in each sample.
7. Construct binding isotherms by fitting data into a Langmuir isotherm to estimate the binding constant (K_b) and number of binding sites (n). Alternative binding models should be used routinely to obtain the best fits to the observed data. Our laboratory often uses the Scatchard analysis or the modified Scatchard analysis to obtain the binding parameters (McGhee & Von Hippel, 1974).

Tip: If the α-ZrP solution was not prepared on the same day, it is advised to stir it for 15 min to ensure its homogeneity before using for binding.

2.3.1 Circular Dichroism Studies

The chirality of natural amino acids and their arrangement in the protein secondary structure result in characteristic circular dichroism (CD) spectra. Hence, the extent of retention of the enzyme secondary structure when bound to α-ZrP is measured by CD spectroscopy. In a typical CD spectrum of protein in the far UV range (190–260 nm), α-helixes are characterized by a double minima at 210 and 222 nm, whereas β-sheets are indicated by a single, broad, negative peak at 212 nm, while random coils are indicated by a sharp negative peak at 195 nm (Pelton & McLean, 2000). However, our goal here is to detect any changes in the enzyme structure upon binding to the nanosheets and hence, we will be comparing the CD spectra before and after binding. Thus, the differential spectra are most sensitive to the enzyme structure changes. The nanosheets themselves have a low absorbance in the deep UV region but thick suspensions of the enzyme/nanosheet complexes scatter light. Therefore, we strongly recommend using very short path lengths, less than 0.5 mm, and most enzymes absorb strongly in the deep UV region of interest and hence, strong signals are obtained even at few micromolar enzyme concentrations. Additionally, we also recommend recording control spectra, as described below, to ensure that the measured CD spectra of the enzyme/nanosheet complexes are not distorted by light scattering. In most of our studies, the CD spectra of the bound enzyme did not indicate any significant alterations and seem to have significant retention of native-like structure.

1. Resuspend the protein/α-ZrP pellet obtained from step 5 above in 1 mL of 10 mM Na$_2$HPO$_4$ buffer at pH 7 and dilute (typically 2 μM) samples, as needed for the CD measurements.

    Tip: Keep the absorbance of the sample to be less than 0.5 in the cuvette to be used for the CD studies.

2. Collect the CD spectra (JASCO J-710 CD spectrophotometer, Easton, MD) of the samples in a 0.05-cm path length quartz cuvette or less, at a scan rate of 10 nm/min with the step resolution of 0.1 nm/data point while setting the bandwidth and sensitivity at 1 nm and 50 millidegree, respectively. Follow the instructions of the manufacturer.

3. Accumulate three scans for each sample and record the average spectrum for each sample, and accumulate more scans until the desired ratio of signal to noise is achieved. Most samples under above conditions gave us good quality spectra, but more scans may be required for some samples.

4. Normalize all spectra per 1 μM of bound protein per 1 cm path length and compare spectra with that of the corresponding unbound enzyme CD spectrum, measured under the same settings.

    Tip: Plot the sample and control curves on the same panel to identify even minor spectral changes due to binding to the nanosheets.

2.3.1.1 Checking for Baseline Correction

Follow the above general method with control samples to test if the solid suspensions by themselves contribute any artifacts to the observed CD spectra. Small path length cuvettes (0.05 cm) give better CD spectra by reducing any scattering interference from the α-ZrP particles. The background scatter from the samples, even though very small, is corrected by placing two identical cells, one containing the exfoliated α-ZrP and the other containing the enzyme solution at exactly the same concentrations and path lengths of the actual samples. The CD spectrum of the double control where cuvette with nanosheets is placed in front of enzyme solution without the nanosheets, is compared with that of the enzyme solution in the absence of nanosheets, and any attenuation in the signal is used to correct for the observed CD spectrum. In our laboratories, this is routinely checked, and we found no significant distortions in the CD spectra of free enzyme solutions when nanosheets suspensions are interposed in front of the sample due to light scattering. Therefore, keeping path lengths to less than or equal to 0.05 cm and keeping the concentrations of the solid suspensions to a minimum are important factors to obtain good quality CD data in the UV CD region of the samples. Even so, each sample needs to be evaluated for any artifacts before proceeding to subsequent studies.

2.3.2 Activity Studies

Catalytic activity provides an insight into the behavior of the bound enzyme with respect to the catalytic activity of the free enzyme in the solution. This comparison is of utmost importance, as these could be used as biocatalysts where the primary role is their enzymatic activity. The accessibility of bound enzyme active site to the substrate and the ease of diffusion of reagents in an out of the galleries of α-ZrP are assessed to maintain enzymatic activities of these heterogeneous samples.

The following protocols are used for the activity studies of several different enzymes used in our studies.

a. Peroxidase activity of Myoglobin (Mb) and Hemoglobin (Hb).

In this colorimetric assay, oxidation of *2-methoxyphenol* (guaiacol) by myoglobin (Mb) or hemoglobin (Hb) in the presence of hydrogen peroxide to form a colored dimeric product ($\lambda_{max} = 470$ nm) is monitored (Scheme 3). The product formation is measured at 470 nm as a function of time, using the in-built kinetic mode in the spectrophotometer (Agilent).

1. Prepare enzyme/α-ZrP as described in the binding studies.
2. Resuspend the enzyme/α-ZrP pellet and dilute the sample with 10 mM Na_2HPO_4 buffer at pH 7.0 in such a way that the final enzyme concentration in the test reaction mixture (2 mL) is 2 μM. In control studies, one will use the freely soluble enzyme solution.
3. Prepare a stock solution of guaiacol in DI water by sonicating the mixture until it completely dissolves.

    Tip: Solubility of guaiacol in water is about 17 mg/mL (15 °C).
4. Add freshly prepared guaiacol (5 mM) solution into the test sample in step 2 in a cuvette, while stirring slowly and constantly.
5. Prepare stock solutions of hydrogen peroxide in DI water.

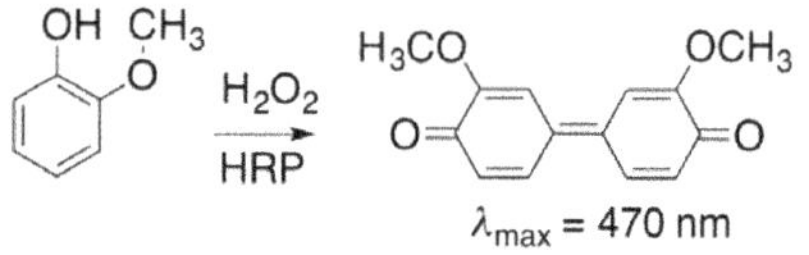

Scheme 3 Oxidase activity of horseradish peroxidase (HRP), and the heme enzymes such as Hb and Mb also catalyze this reaction. The enzymatic activity is measured using o-methoxyphenol as the substrate while monitoring product formation at 470 nm.

Tip: The stock solutions of hydrogen peroxide are around 30% and strongly corrosive. Take precautions and dilute these to a lower concentration before handling.
6. Set spectrophotometer to kinetic mode sampling at 470 nm calculating initial rate every second from 5 to 60 s.
7. Blank the spectrophotometer with 10 mM sodium phosphate buffer pH 7.0.
8. Mix hydrogen peroxide (1 mM) into the reaction mixture in step 4, while recording the absorbance of the sample for 1 min or more.
    Tip: If the reaction is faster, use a stopped flow apparatus. Collect 20–50 points in the initial time window of the kinetics, which need to be linear for proper analysis.
9. Graph kinetic data in terms of absorbance at 470 nm versus time (Kaleidagraph, version 4.1.3 or equivalent) and calculate the initial rates from the initial slope of the graph.

b. Lysozyme (Lys) activity assay.

Hydrolytic activity of lysozyme is monitored by the hydrolysis of glycol chitin, a soluble, chemically modified form of chitin, a known substrate for lysozyme. The production of the reducing end groups by the enzymatic action on chitin is monitored by oxidation with potassium ferricyanide at 420 nm.

1. Incubate lysozyme (60 μM) or lysozyme bound to the nanosheets with increasing concentrations of the substrate, glycol chitin, 0.05% to 0.3 wt.% at 40 °C for 1.25 h.
2. Add potassium ferricyanide (2 mL, 15 mM in 0.5 M sodium carbonate) to the above solutions in step 1 and heat the samples in boiling water for 30 min to complete the oxidation of the reductive end groups generated by the enzymatic action.
3. Cool down samples to room temperature and measure the absorbance at 420 nm due to potassium ferrocyanide.
4. Prepare a standard calibration curve of absorbance versus product concentration by mixing known concentrations of ferricyanide solution and n-acetylglucosamine.
5. Estimate the lysozyme activity using the above standard curve.
6. Follow the same protocol for lysozyme/α-ZrP (7 μM lysozyme, 1.5 mM α-ZrP) and compare the initial rates.

c. Chymotrypsin (CT) activity assay.

Assay for CT is based on the hydrolysis of benzoyl-L-tyrosine ethyl ester (BTEE) by the enzyme, and the increase in absorbance at 256 nm corresponds to the formation of N-benzoyl-L-tyrosine (Scheme 4).

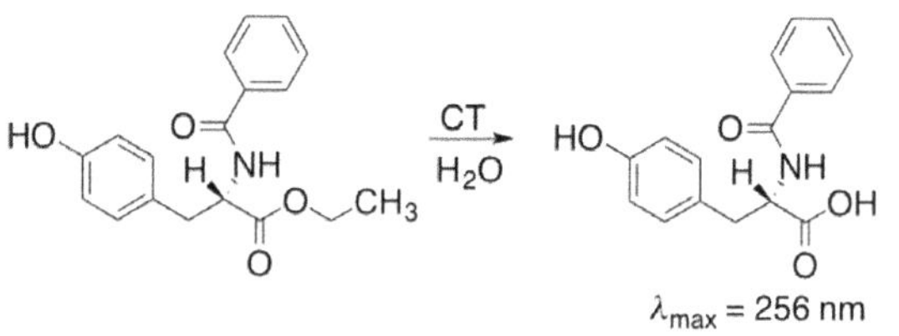

Scheme 4 Hydrolysis of N-benzyl-L-tyrosine ethyl ester by CT to form N-benzoyl-L-tyrosine.

1. Prepare stock solution of CT in 20 mM Tris buffer containing CaCl$_2$ (38 mM) and methanol (25%) at pH 7.8.
2. Resuspend CT/α-ZrP sample in the above buffer and mix CT/α-ZrP (1 μM) with BTEE.
3. Record the absorbance of samples at 256 nm at the intervals of 1, 12, and 24 h in a quartz cuvette.

   Tip: Do not use glass cuvettes as these absorb strongly at the above wavelength.
4. Calculate the rates from difference in absorbance (ΔA) of the product and the starting materials using the known extinction coefficient of BTEE (0.964 mM/cm) at 256 nm (Kumar & Chaudhari, 2002).

2.4 Binding of Negatively Charged Proteins to Anionic α-ZrP Nanosheets

Though the phosphate lattice of α-ZrP provides a uniform, hydrated charged surface for interactions with the positively charged enzymes and proteins, the strong negative charge field limits the binding of negatively charged biomolecules due to the unfavorable electrostatic repulsions. As a result, tuning of the surface charge of the solid is required prior to the protein binding.

2.4.1 Metal Ion Mediated Protein Binding to the Nanosheets

One approach to neutralize the excess negative charge that would accumulate when anionic enzymes bind to anionic nanosheets is to use excess metal ions in the buffer for charge neutralization (Fig. 3). Sequestration of metal ions at the interface neutralizes the excess negative charges at the interface and hence improves the binding of negatively charged proteins to the negatively charged solid surface. Here, we provide a simple protocol for metal-mediated protein binding to α-ZrP nanosheets (Fig. 3).

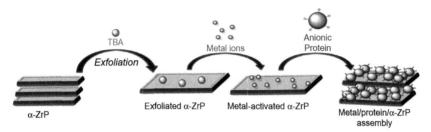

Figure 3 Cation-mediated binding of anionic enzymes to anionic α-ZrP nanosheets. (See the color plate.)

1. Add increasing concentrations of metal ions (e.g., Zr^{4+}, 0–1.5 mM, using ZrOCl$_4$ as the source of Zr^{4+} at pH 3 so that there is no precipitation of the protein or the metal hydroxide formation) to constant concentration of negatively charged protein (e.g., glucose oxidase (GOx), 20 μM) and α-ZrP (3 mM) in DI water.

    Tip: Other metal ions of +2, +3, or +4 charge may be substituted for Zr(IV), but the corresponding buffer ions and pH are to be maintained so that there is no precipitation of the metal hydroxide. Thus, the solubility product and the isoelectric point of the metal hydroxide need to be kept in mind for this experiment.

2. Check for any precipitation of protein in the presence of metal ions and pick the concentration of the metal ion, pH, and buffer ions such that there is no metal or enzyme precipitation under these conditions.

3. Add increasing concentrations of enzyme in the presence of metal ions of interest at the optimum concentration range (e.g., Zr^{4+}, 1 mM) and α-ZrP (3 mM).

4. Equilibrate the enzyme/metal ion/α-ZrP mixtures for 30 min at room temperature.

5. Separate the free enzyme from that of the bound one by centrifuging the samples at 10,000 rpm for 5 min (Fisher Scientific Microfuge, Accuspin 17).

6. Determine the concentration of free proteins in the supernatants by monitoring the absorbance at 280 nm using the corresponding extinction coefficient (e.g., ε_{280nm}, GOx $= 1.336 \times 10^5\ M^{-1}\ cm^{-1}$ (Pattammattel, Deshapriya, et al., 2013)).

7. Construct binding isotherms from the data from step 6 to determine the binding constant and the number of binding sites.

    Tip: The order of addition of the components is as follows: protein, DI water, metal ion, and α-ZrP suspension.

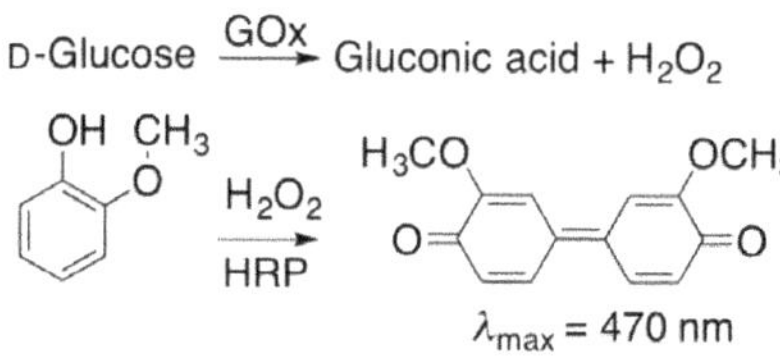

$$\text{D-Glucose} \xrightarrow{\text{GOx}} \text{Gluconic acid} + H_2O_2$$

[Guaiacol oxidation scheme: guaiacol (OH, OCH₃ on benzene) + H_2O_2 → tetraguaiacol (H₃CO, OCH₃ substituted bis-quinone), catalyzed by HRP; $\lambda_{max} = 470$ nm]

Scheme 5 Production of H_2O_2 during the oxidation of D-glucose to gluconic acid by oxygen, catalyzed by GOx.

2.4.2 Activity Assay for GOx/Zr^{4+}/α-ZrP

The assay indirectly measures the amount of hydrogen peroxide produced during the oxidation of glucose to gluconic acid by GOx using the ambient oxygen (Scheme 5).

1. Prepare GOx/Zr^{4+}/α-ZrP as indicated in the above method.
2. Prepare stock solutions of guaiacol, horseradish peroxidase (HRP), and glucose in DI water.
3. Add solutions of guaiacol (5 m*M*) and HRP (1 µ*M*) into the solution in step 1, under continuous stirring.
4. Set spectrophotometer to kinetic mode sampling at 470 nm calculating initial rate every second from 5 to 60 s or longer as needed.
5. Blank the spectrophotometer with 10 m*M* sodium phosphate buffer pH 7.
6. Immediately mix glucose (2 m*M*) into the reaction mixture while recording the absorbance for 1 min.
7. Graph kinetic data in terms of absorbance at 470 nm versus time (Kaleidagraph, version 4.1.3 or equivalent) and calculate the initial rates from the slope of the graph.

2.4.3 Use of Protein Glues for Enzyme Binding

An alternative and attractive way to overcome the electrostatic barrier to bind anionic enzymes to anionic nanosheets is to use a suitable protein coating, which adjusts the surface charge of the nanosheets as desired. One additional advantage of this method is that the nanosheet surface is also coated with a proteinaceous material such that the surface is friendlier to biological systems.

The protein coating on the nanosheet is chosen such that it binds to both the solid and the enzyme with high affinities. To produce a highly positively charged protein coating, BSA is chemically modified to form cationized BSA. Aspartic and glutamic acid residues of BSA (pI 4.7) are activated with carbodiimide ethyl-3-(3-(dimethylamino)propyl)carbodiimide (EDC) followed by the reaction with polyamine (triethylenetetramine, TETA) to form the corresponding BSA–TETA derivative. Each TETA

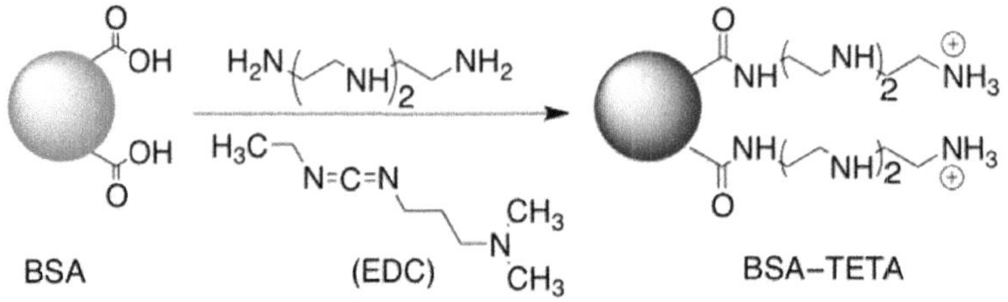

Scheme 6 Cationization of bovine serum albumin (BSA) with triethylenetetramine (TETA) using carbodiimide chemistry.

side chain can provide multiple positive charges, depending on the pH conditions (Scheme 6).

The following protocol gives a BSA–TETA (also named as cBSA in the text) derivative with an overall charge of +27, as measured by using electrophoretic mobility in standard agarose gels, and the charge is estimated relative to that of BSA which has a known charge of −18, under the experimental conditions. The magnitude of the charge on the modified BSA sample is controlled by adjusting the experimental conditions described below.

1. Add BSA solution (1.5 g, 8 mL DI water) into a solution of TETA (0.5 M, 5 mL DI water).
2. Adjust the pH to 5.0 using concentrated HCl and stir the solution for 15 min at room temperature.

   Tip: Concentrated hydrochloric acid is highly corrosive and use proper safety procedures in handling the reagent.
3. Add EDC (1 M, 1 mL in DI water) and stir for 4 h.
4. Dialyze the sample against 10 mM sodium phosphate buffer at pH 7.0 using 25 kDa cutoff membrane for three times.

Binding of negatively charged proteins to exfoliated α-ZrP requires significant charge neutralization by cBSA. cBSA binding results in charge neutralization followed by charge reversal (∼+40 mV), which can easily be monitored by a zeta potential titration. Hence, the positively charged α-ZrP (named as bZrP), with the charge of +40 mV, is used for negatively charged enzyme binding and subsequent experiments. Absorption of BSA–TETA (+27) to exfoliated α-ZrP is achieved by simple mixing of two reagents under predetermined concentrations to form positively charged bZrP derivative with the desired charge. Charge and the morphology of bZrP can be tested using zeta potential and EDAX measurement. bZrP is then mixed with the negatively charged protein of interest to form enzyme/bZrP complexes as described before (Fig. 4).

Figure 4 Binding of negatively charged enzymes/protein into the galleries of exfoliated α-ZrP with the aid of cationized BSA. (See the color plate.)

The affinity of strongly acidic enzymes at pH 7.0 (e.g., pepsin, GOx, tyrosinase, laccase) is tested for bZrP by binding studies following the same protocol used in Section 2.4.1 by replacing α-ZrP with bZrP. Enzyme/bZrP complexes are further characterized for bound enzyme structure, activity, and stabilities.

2.4.4 Preparation of Enzyme/bZrP Complexes

1. Add 30 μM cBSA to exfoliated α-ZrP (6 mM, 10 mM sodium phosphate buffer, pH 7) while mixing carefully.
2. Add enzyme of interest into the above mixture in step 1 equilibrate for 1 h at room temperature.
3. Monitor the extent of binding by centrifugation studies discussed before. The secondary structure of the acidic enzymes bound to bZrP can be monitored by CD spectra as described before, whereas enzymatic activity assays for the above selected set of enzymes are performed by following detailed procedures in this chapter.

a. Assay for pepsin.

Pepsin activity assay is performed with Hb as the substrate. Hb is first denatured by trichloroacetic acid (TCA) and the rate of hydrolysis is measured.

   1. Dissolve pepsin/bZrP (0.5 mg/mL of enzyme) in 0.01 N HCl and cool down to room temperature (25 °C) prior to the assay.
   2. Dissolve Hb (2.5 g) in 100 mL DI water and dilute 80 mL of this solution with 20 mL of 0.3 N HCl.
   3. Add above solution in step 2 (2.5 mL each) into nine numbered test tubes and incubate at 37 °C in a water bath.
   4. Keep tubes 1–3 as blanks and add 5 mL of TCA (5%, w/v) solution into each tube followed by pepsin (2 mL).
   5. Take samples out after 5 min, filter, and measure the absorbance at 280 nm.
   6. Add pepsin/bZrP (2 μM) to test tubes 4–6 at timed intervals and equilibrate the mixtures at 37 °C for 10 min.

7. Add TCA (5 mL) into each tube to quench the reaction and after 5 min, filter, and measure the absorbance at 280 nm.
8. Similarly treat samples 7–9 with unbound pepsin (2 µM).
9. Subtract the absorbance of the blank at 280 nm from each test sample to get the extent of hydrolysis by pepsin.

b. Assay for tyrosinase.

Oxidation of L-tyrosine to dihydroxyphenylalanine which inturn is oxidized to o-quinone is measured during the assay. Formation of L-DOPA (4-dihydroxyphenylalanine)-quinone causes increase in absorbance at 280 nm (Scheme 7).

1. Dissolve L-tyrosine (20 µM) in 0.5 M phosphate buffer at pH 6.5.
2. Set spectrophotometer to kinetic mode sampling at 280 nm calculating initial rate every second from 5 to 60 s.
3. Blank the spectrophotometer with 0.5 M sodium phosphate buffer pH 6.5.
4. Immediately mix tyrosinase/bZrP (2 µM) in the same buffer into the above solution recording the absorbance for 1 min.
5. Graph kinetic data in terms of absorbance at 470 nm versus time (Kaleidagraph, version 4.1.3) and calculate the initial rates from the slope of the graph.

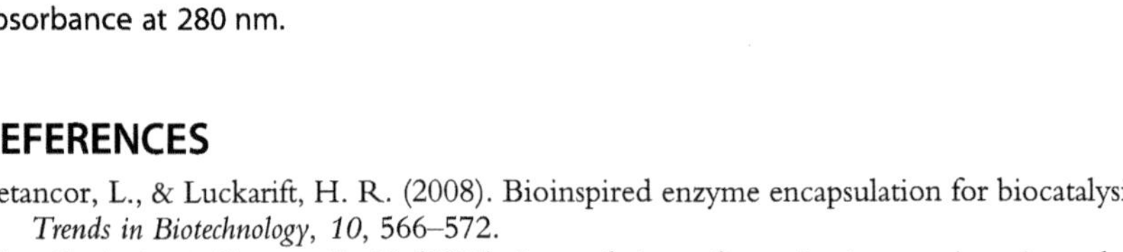

L-Tyrosine → L-DOPA (λ = 280 nm) → L-DOPA-quinone

Scheme 7 Conversion of L-tyrosine to L-DOPA by tyrosinase is measured by increasing absorbance at 280 nm.

REFERENCES

Betancor, L., & Luckarift, H. R. (2008). Bioinspired enzyme encapsulation for biocatalysis. *Trends in Biotechnology, 10*, 566–572.

Chaudhari, A., & Kumar, C. V. (2005). Intercalation of proteins into α-zirconium phosphonates: Tuning the binding affinities with phosphonate functions. *Microporous and Mesoporous Materials, 2–3*, 175–187.

Clearfield, A., & Smith, G. D. (1969). Crystallography and structure of α-zirconium bis(monohydrogen orthophosphate) monohydrate. *Inorganic Chemistry, 3*, 431–436.

Cole-Hamilton, D. J. (2003). Homogeneous catalysis—New approaches to catalyst separation, recovery, and recycling. *Science, 5613*, 1702–1706.

Costantino, H. R., Firouzabadian, L., Hogeland, K., Wu, C., Beganski, C., Carrasquillo, K. G., et al. (2000). Protein spray-freeze drying. Effect of atomization conditions on particle size and stability. *Pharmaceutical Research, 17*(11), 1374–1383.

Deshapriya, I. K., Kim, C. S., Novak, M. J., & Kumar, C. V. (2014). Biofunctionalization of α-zirconium phosphate nanosheets: Toward rational control of enzyme loading, affinities, activities and structure retention. *ACS Applied Materials & Interfaces, 12*, 9643–9653.

DiCosimo, R., McAuliffe, J., Poulose, A. J., & Bohlmann, G. (2013). Industrial use of immobilized enzymes. *Chemical Society Reviews, 15*, 6437–6474.

Duff, M. R., & Kumar, C. V. (2009a). Molecular signatures of enzyme–solid interactions: Thermodynamics of protein binding to α-Zr(IV) phosphate nanoplates. *The Journal of Physical Chemistry B, 45*, 15083–15089.

Duff, M. R., & Kumar, C. V. (2009b). Protein-solid interactions: Important role of solvent, ions, temperature, and buffer in protein binding to α-Zr(IV) phosphate. *Langmuir, 21*, 12635–12643.

Garcia, M. E., Naffin, J. L., Deng, N., & Mallouk, T. E. (1995). Preparative-scale separation of enantiomers using intercalated alpha-zirconium phosphate. *Chemistry of Materials, 10*, 1968–1973.

Kaschak, D. M., Johnson, S. A., Hooks, D. E., Kim, H.-N., Ward, M. D., & Mallouk, T. E. (1998). Chemistry on the edge: A microscopic analysis of the intercalation, exfoliation, edge functionalization, and monolayer surface tiling reactions of α-zirconium phosphate. *Journal of the American Chemical Society, 42*, 10887–10894.

Kim, H.-N., Keller, S. W., Mallouk, T. E., Schmitt, J., & Decher, G. (1997). Characterization of zirconium phosphate/polycation thin films grown by sequential adsorption reactions. *Chemistry of Materials, 6*, 1414–1421.

Kumar, C. V., & McLendon, G. L. (1997). Nanoencapsulation of cytochrome c and horseradish peroxidase at the galleries of -zirconium phosphate. *Chemistry of Materials, 9*, 863.

Kumar, C. V., & Chaudhari, A. (2000). Proteins immobilized at the galleries of layered α-zirconium phosphate: Structure and activity studies. *Journal of the American Chemical Society, 5*, 830–837.

Kumar, C. V., & Chaudhari, A. (2002). High temperature peroxidase activities of HRP and hemoglobin in the galleries of layered Zr(IV)phosphate. *Chemical Communications, 20*, 2382–2383.

Lu, H., Wilkie, C. A., Ding, M., & Song, L. (2011). Thermal properties and flammability performance of poly(vinyl alcohol)/α-zirconium phosphate nanocomposites. *Polymer Degradation and Stability, 5*, 885–891.

Mateo, C., Palomo, J. M., Fernandez-Lorente, G., Guisan, J. M., & Fernandez-Lafuente, R. (2007). Improvement of enzyme activity, stability and selectivity via immobilization techniques. *Enzyme and Microbial Technology, 6*, 1451–1463.

McGhee, J. D., & von Hippel, P. H. (1974). Theoretical aspects of DNA-protein interactions: co-operative and non-co-operative binding of ligands to a one-dimensional homogeneous lattice. *Journal of Molecular Biology, 86*, 469–489.

Mudhivarthi, V. K., Bhambhani, A., & Kumar, C. V. (2007). Novel enzyme/DNA/inorganic nanomaterials: A new generation of biocatalysts. *Dalton Transactions, 47*, 5483–5497.

Pattammattel, A., Deshapriya, I. K., Chowdhury, R., & Kumar, C. V. (2013). Metal-enzyme frameworks: Role of metal Ions in promoting enzyme self-assembly on α-zirconium(IV) phosphate nanoplates. *Langmuir, 9*, 2971–2981.

Pattammattel, A., Puglia, M., Chakraborty, S., Deshapriya, I. K., Dutta, P. K., & Kumar, C. V. (2013). Tuning the activities and structures of enzymes bound to graphene oxide with a protein glue. *Langmuir, 50*, 15643–15654.

Pelton, J. T., & McLean, L. R. (2000). Spectroscopic methods for analysis of protein secondary structure. *Analytical Biochemistry, 2*, 167–176.

Puri, M., Barrow, C. J., & Verma, M. L. (2013). Enzyme immobilization on nanomaterials for biofuel production. *Trends in Biotechnology, 4*, 215–216.

Schmid, A., Dordick, J. S., Hauer, B., Kiener, A., Wubbolts, M., & Witholt, B. (2001). Industrial biocatalysis today and tomorrow. *Nature, 6817*, 258–268.

Schoemaker, H. E., Mink, D., & Wubbolts, M. G. (2003). Dispelling the myths—Biocatalysis in industrial synthesis. *Science, 5613*, 1694–1697.

CHAPTER NINE

Portable Enzyme-Paper Biosensors Based on Redox-Active CeO$_2$ Nanoparticles

A. Karimi, A. Othman, S. Andreescu[1]

Department of Chemistry and Biomolecular Science, Clarkson University, Potsdam, New York, USA
[1]Corresponding author: e-mail address: eandrees@clarkson.edu

Contents

1. Introduction — 178
2. NPs-Based Enzyme Biosensors — 179
   2.1 NPs for Biosensing Design — 179
   2.2 Immobilization of Enzymes on NPs — 180
   2.3 Fabrication of Low-Cost Enzyme Biosensors — 181
3. CeO$_2$ NPs for Enzyme Immobilization and Enzyme-Based Biosensors — 183
4. Design of a CeO$_2$-Based Colorimetric Enzyme Biosensor — 185
   4.1 Method Principle and Main Characteristics — 185
   4.2 Materials — 186
   4.3 Methods — 187
   4.4 Detection and Measurement Procedure — 188
5. Comments on the Method — 190
   5.1 NP Choice — 190
   5.2 Enzyme and NP Immobilization — 191
   5.3 Testing for Interferences — 191
   5.4 Color Analysis — 191
Acknowledgments — 192
References — 192

Abstract

Portable, nanoparticle (NP)-enhanced enzyme sensors have emerged as powerful devices for qualitative and quantitative analysis of a variety of analytes for biomedicine, environmental applications, and pharmaceutical fields. This chapter describes a method for the fabrication of a portable, paper-based, inexpensive, robust enzyme biosensor for the detection of substrates of oxidase enzymes. The method utilizes redox-active NPs of cerium oxide (CeO$_2$) as a sensing platform which produces color in response to H$_2$O$_2$ generated by the action of oxidase enzymes on their corresponding substrates. This avoids the use of peroxidases which are routinely used in conjunction with glucose oxidase. The CeO$_2$ particles serve dual roles, as high surface area supports to anchor high loadings of the enzyme as well as a color generation reagent, and the particles are recycled multiple times for the reuse of the biosensor. These sensors are small, light, disposable, inexpensive, and they can be mass produced by standard, low-cost printing methods. All reagents needed for the analysis are embedded within the paper matrix, and sensors stored over extended periods of time without performance loss. This novel sensor is a general platform for the in-field detection of analytes that are substrates for oxidase enzymes in clinical, food, and environmental samples.

ABBREVIATIONS
APTS aminopropyltrimethoxysilane
CeO$_2$ cerium oxide
GOX glucose oxidase
NPs nanoparticles

Paper
Hydrogen peroxide

1. INTRODUCTION

Successful implementation of enzyme-based (bio)analytical systems and devices requires the development of functional enzyme-based materials and surfaces that retain the activity and function of the enzymes over substantial time periods, under ordinary conditions. In recent years there has been significant research on the use of nanostructured materials, nanoparticles (NPs) and nanocomposites to develop functional enzyme-based devices. Enzyme attachment and assembly of enzyme–NP conjugates are critical steps in the fabrication of enzyme-based biosensors. When used in conjunction with enzymes, nanomaterials can increase enzyme stability and enhance binding efficiency leading to improved performance of sensors. Enzyme–NP conjugates can be used as building blocks for the development of inexpensive, portable, and easy-to-use biosensing devices. The interesting feature of the current method is that the NPs serve as supports for the enzyme binding and also serve as the colorimetric reagent for analyte detection. The advances in the development, fabrication, characterization, and implementation of functional enzyme–NP conjugate systems for field-portable biosensors are described here. Due to their light weight and simple-to-use features, they are suitable for the on-site detection of a variety of analytes of interest for clinical diagnosis, environmental monitoring, public safety, and food quality control.

The demand for reliable, field-deployable and cost-effective devices that can provide rapid on-site assessment of the environment, water, air, or food chain continues to grow (Ding, Srinivasan, & Tung, 2015). The availability of affordable technology can enable local communities in resource limited and remote regions to improve health care, protect the environment, and maintain the safety of food. Such sensors can rapidly facilitate the World Health Organization's (WHO) goal for *Affordable, Sensitive, Specific, User-friendly, Rapid and Robust, Equipment free, and Deliverable Sensors to end-users*, termed as "ASSURED" (Mabey, Peeling, Ustianowski, & Perkins, 2004). Enzyme-based methods have gained increasing acceptance in the biomedical field (Ispas, Crivat, & Andreescu, 2012; Wilson & Hu, 2000) and their use could be extended to other application areas as well. Enzyme-based devices combine the selective recognition capabilities of enzymes with the optoelectronic, catalytic, and mechanical properties of nanomaterials. Interfacing enzymes with nanostructured materials can generate functional enzyme–nanomaterial conjugates that can be used as active components in the development of practical enzyme-based assays for field monitoring. Nanomaterials have been shown to increase the stability and robustness, sensitivity, and shelf life of enzyme-based devices (Andreescu, Njagi, & Ispas, 2008). With the development of advanced nanomanufacturing techniques, enzyme-based devices that incorporate NPs can be miniaturized, mass produced, and multiplexed. They can be developed for real-time and point-of-care diagnostic testing (Ispas et al., 2012) and used in the field (Gabig-Ciminska, 2006).

The properties, fabrication, and characterization of enzyme–NP conjugates and their application for the development of field-deployable biosensors are described here. After a brief discussion of the current development status of NP-based enzyme biosensors, we describe a method that can be used as a general platform for the immobilization of enzymes and NPs on paper. An example of application for the immobilization of glucose oxidase (GOX) and redox-active NPs of CeO_2 to fabricate a fully integrated portable assay for the colorimetric detection of glucose using NPs as color-generating probes is given here.

2. NPs-BASED ENZYME BIOSENSORS

2.1 NPs for Biosensing Design

Metal and metal oxide NPs are the most common types of NPs used to design portable enzyme-based biosensing platforms. In a traditional enzyme-based biosensor design, NPs can serve as: (i) signal amplifier to increase sensitivity; (ii) immobilization support to stabilize the enzyme; and (iii) as a sensor platform to increase surface area, enhance conductivity in electrochemical assays, or provide a colorimetric read-out in optical sensing schemes. The unique physicochemical properties of the NPs (i.e., catalysis, surface plasmon, magnetism, luminescence, conductivity, etc.) associated with their small size and tailored surface chemistry can generate powerful enzyme–NP conjugate systems that can be used as active components in biosensing devices.

Progress in nanotechnology has provided the ability to tailor the characteristics of NPs through synthetic design by controlling the particle size, surface properties, and chemical composition. Therefore, NPs with enhanced electronic, optical, or catalytic functions can be obtained (Baron, Willner, & Willner, 2007). The NP chemistry, its structure, surface charge, functionality, and composition have a critical role on the binding, biological activity, orientation, structure, and functionality of enzymes (Jo, Kim, Lee, & Kim, 2015; Kim, Grate, & Wang, 2006; Popat et al., 2011). Enzyme function can also be tuned by carefully selecting the characteristics of the medium such as pH, temperature, viscosity, and salt concentrations of the measurement environment (Czeslik & Winter, 2001; Jia, Zhu, & Wang, 2003). Functional enzyme–NP assemblies and biofunctionalized NP composites can find broad applicability in biosensing (Demin & Hall, 2009; Pingarrón, Yáñez-Sedeño, & González-Cortés, 2008), biocatalysis (Schoemaker, Mink, & Wubbolts, 2003), tissue engineering (Arruda et al., 2009), disease diagnosis (Arruda et al., 2009), biofuel cells (Naruse et al., 2011), and drug delivery (Ai, Jones, & Lvov, 2003).

2.2 Immobilization of Enzymes on NPs

Enzyme stabilization on sensor surfaces is an essential step for the development of reagentless and fully integrated field-portable biosensors. This process is governed by various interactions between the enzyme and the material interface. The composition and properties of nanoscale surfaces and interfaces control the functionality of these devices. The main challenge is to design and control the interfacial forces such that it will allow stable enzyme attachment while maintaining activity and function as close as possible to its native state. These materials should either possess the necessary functional groups needed for enzyme attachment or should be easily functionalized. Enzymes immobilized on NPs and NP-based composites have demonstrated increased stability over free enzymes (Homaei, Sariri, Vianello, & Stevanato, 2013; Lata et al., 2015). The use of immobilized enzymes is preferred due to their ease of handling, prolonged availability, and robustness, increased resistance to environmental changes and reusability (Datta, Christena, & Rajaram, 2013).

The NP characteristics such as size and surface chemistry as well as the orientation and density of the immobilized enzymes all contribute to their catalytic performance. Enzyme configuration, orientation, and density can be controlled by changing the NPs surface chemistry (Ardao, Comenge, Benaiges, Álvaro, & Puntes, 2012). These can be tailored to promote electrostatic binding (surface charge) or by adding surface ligands for covalent (e.g., through glutaraldehyde or carbodiimide chemistry) or affinity (e.g., thiol) attachment. Enzyme entrapment within NP-based composite materials with biocompatible or conducting polymers is another strategy commonly used to create enzyme-based bioassays (Homaei et al., 2013). NPs can be used as enzyme carriers (Ding, Cargill, Medintz, & Claussen, 2015) to increase enzyme loading by virtue of their high surface area and reduce substrate diffusion limitations (Won, Aboagye, Jang, Jitianu, & Stanciu, 2010). In electrochemical assays, NPs can promote electronic communication between redox enzymes and electrode surfaces, which is essential for the development of electrochemical enzyme-based biosensors (Heller, 1990). Figure 1 shows a summary of the various NPs systems and their applications in enzyme-based biosensors.

2.3 Fabrication of Low-Cost Enzyme Biosensors

Development of low-cost and portable devices for home diagnostics, screening and point-of-care testing requires fabrication and assembly of biofunctionalized surfaces on inexpensive substrates that have the appropriate detection chemistry, sensitivity, and selectivity. A key requirement in the production of enzyme-based bionanofunctionalized structures for portable and low-cost sensing is to achieve uniform deposition of NP–enzyme conjugates to ensure the required sensitivity and preserve the biological activity of the enzyme. In portable platforms, NPs are often embedded in polymeric films and are prone to aggregation, losing their nanoscale features. As a result, there is little control over the nanoassembly, and both NPs and enzymes are randomly distributed within the nanostructured network, which strongly affect the sensitivity and reproducibility of sensing devices. Many enzyme biosensors reported in literature involve complicated multistep procedures,

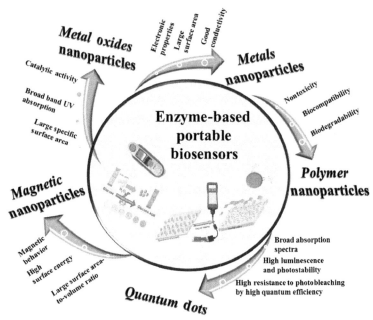

Figure 1 Overview of the various NPs systems and their properties for enzyme-based portable biosensors. (See the color plate.)

expensive substrates and reagents, most of which are carried out manually, and require special conditions to conserve bioactivity. In many cases reagents needed for detection or signal amplification (including bioreagents, colorimetric dyes, and nanoelements) are added postfabrication, adding additional cost, and inconvenience for analysis. Automatic deposition procedures such as inkjet and 3D printing enable scalable production with increased reproducibility (Abe, Suzuki, & Citterio, 2008; Gowers et al., 2015).

Low-cost biosensing devices have been developed recently on paper, thread, or plastic as alternatives to conventional instrumentation for field analysis (Martinez, Phillips, Butte, & Whitesides, 2007; Metters, Houssein, Kampouris, & Banks, 2013; Riccardi et al., 2016). Applications include point-of-care medical diagnosis, environmental monitoring, and food quality control (Martinez et al., 2007; Martinez, Phillips, Whitesides, & Carrilho, 2010; Pelton, 2009; Zhao, Ali, Aguirre, Brook, & Li, 2008). These platforms have been integrated with colorimetric (Dungchai, Chailapakul, & Henry, 2010; Martinez et al., 2007) and electrochemical (Dungchai, Chailapakul, & Henry, 2009; Nie et al., 2010; Reches et al., 2010) signal transduction systems. Such devices can be miniaturized, they are disposable and can be used for on-site analysis. Some examples of paper-based bioassays include: patterned-paper fabricated by photolithography for the detection of glucose and bovine serum albumin (Martinez et al., 2007), sol–gel bioinks for the detection of neurotoxins (Hossain et al., 2009), and aptamer—NP-based lateral flow devices for the detection of DNA sequences (Liu, Mazumdar, & Lu, 2006). Several NP-based bioassays have been reported; few of these have been adapted for implementation with paper platforms (Zhao, Brook, & Li, 2008). Traditionally, these assays are based on NP aggregation or dispersion of colloidal gold in the presence of the analyte (Elghanian, Storhoff, Mucic, Letsinger, & Mirkin, 1997; Zhao, Brook, et al., 2008). An example is the detection of glucose in urine using colloidal Au NPs functionalized with GOX (Radhakumary, 2011). These assays utilize presynthesized NPs, and in vast majority of the cases, these are used in colloidal dispersions. Methods that enable large scale fabrication of bionanofunctionalized surfaces that conserve both the nanoscale properties and bioactivity are of great interest for the manufacturing of inexpensive, portable sensors. Current research focuses on the use of 3D and inkjet technology for enzyme printing. These methods can potentially be used for the automatic fabrication of the entire biosensing device.

An enzyme-based NP system that takes advantage of the catalytic, optical, and redox properties of redox-active CeO_2 NPs is described here. We couple the optical properties of these NPs with the versatile and inexpensive nature of paper and the catalytic activities of redox enzymes to design a portable enzyme-based biosensor.

3. CeO_2 NPs FOR ENZYME IMMOBILIZATION AND ENZYME-BASED BIOSENSORS

CeO_2 NPs have demonstrated interesting properties for sensing, biomedical, and drug delivery applications (Andreescu et al., 2014). These NPs are unique because of the presence of cerium in dual oxidation states (Ce^{3+}/Ce^{4+}) at their surface, which enable them to be used as both oxidation and reduction catalysts (Pirmohamed et al., 2010). CeO_2 NPs have high oxygen mobility at their surface (Preda et al., 2011; Zhang, Wang, Wang, & Li, 2006) and a large oxygen diffusion coefficient, which facilitates the conversion between valance states Ce^{3+}/Ce^{4+} and allows oxygen to be released or stored in the crystal structure (Dutta et al., 2006; Wang et al., 2011;

Xu et al., 2010). Their catalytic, oxygen rich properties, rich functional groups, and low toxicity (Das et al., 2007) make these NPs an excellent immobilization material for enzymes (Ansari, Solanki, & Malhotra, 2009; Ornatska, Sharpe, Andreescu, & Andreescu, 2011). CeO_2 NPs have unique optical properties derived from the redox conversion between the two oxidation states and can form charge transfer complexes of unique colors with H_2O_2 and with phenol-type compounds. The optical and redox properties of these NPs enable them to be used as colorimetric indicators to replace commonly used soluble dyes (Ornatska et al., 2011; Sharpe et al., 2014; Sharpe, Frasco, Andreescu, & Andreescu, 2013) which provide exciting opportunities for the rational design of portable and robust enzyme/NP/biosensors.

The use of CeO_2 NPs in contact or proximity with oxidase enzymes can eliminate or minimize problems associated with low oxygen levels for enzymes that require oxygen as a cosubstrate, enabling these biosensors to work in hypoxic environments. The catalytic and oxygen storage/release capacity of CeO_2 NPs were used to fabricate "oxygen rich" electrochemical biosensors (Ozel, Ispas, Ganesana, Leiter, & Andreescu, 2014). CeO_2 also facilitates the stabilization of the enzyme through strong electrostatic interactions, thereby enhancing the operational and storage stability of enzyme sensors based on these materials (Khan & Dhayal, 2008; Li et al., 2006; Njagi, Ispas, & Andreescu, 2008; Topoglidis, Cass, O'Regan, & Durrant, 2001). Coimmobilization of oxidase enzymes with these particles greatly improved performance of *in vivo* biosensors under hypoxic conditions for real-time measurements of dopamine with tyrosinase (Njagi, Chernov, Leiter, & Andreescu, 2010), lactate with lactate oxidase (Sardesai, Ganesana, Karimi, Leiter, & Andreescu, 2015), and glutamate with glutamate oxidase (Ozel et al., 2014). Enhanced capabilities were achieved by doping the CeO_2 particles with Pt to amplify the catalytic activity and imparting conductivity, which significantly increased the detection sensitivity. For example, a Pt–CeO_2-based biosensor fabricated by coimmobilizing the particles with GOX in a porous chitosan matrix enabled sensitive detection of physiologically relevant glucose concentrations in conditions of severe oxygen depletion, with a detection limit of 37.4 μM and of lactate in implantable conditions (Sardesai et al., 2015; Sardesai, Karimi, & Andreescu, 2014). In vitro characterization of the biosensor demonstrated high selectivity against physiological levels of ascorbic acid, a storage stability of 3 weeks, a response time of 6 s, and linear sensitivity over a wide concentration range. Other applications of this system include implantable enzymatic biofuel cells and wearable devices.

4. DESIGN OF A CeO₂-BASED COLORIMETRIC ENZYME BIOSENSOR

4.1 Method Principle and Main Characteristics

This method takes advantage of the color changes resulting from the alterations in the redox state and composition of NPs in response to H_2O_2, the product of the enzymatic reaction of most oxidase enzymes. CeO_2 NPs and the enzyme are both incorporated into a paper-based format. Ceria NPs have been found to possess oxidase and peroxidase-like activity (Asati, Santra, Kaittanis, Nath, & Perez, 2009), suggesting that they could potentially replace oxidase and peroxidase enzymes (Ornatska et al., 2011). The coexistence of two oxidation states imparts unique surface chemistry, redox, and catalytic activity (Davis & Thompson, 2002; Zhang, Wang, Koberstein, Khalid, & Chan, 2004) which make these particles interesting candidates for biosensing applications. The method described here is demonstrated for the detection of glucose using GOX as a model system (Ornatska et al., 2011) but the same approach can be used to fabricate other oxidase enzyme-based biosensors. To fabricate the biosensor, CeO_2 NPs, and GOX (binds specifically to β-D-glucopyranose and does not act on α-D-glucose) are coimmobilized onto filter paper using a silanization procedure. In the presence of glucose, the enzymatically generated H_2O_2 induces a visual color change of the CeO_2 NPs, immobilized onto the modified paper surface, from white-yellowish to dark-orange, in a concentration-dependent manner. The visible color change is due to the change in the oxidation state and formation of surface complexes (Babko, 1954; Yu, Hayes, O'Keefe, O'Keefe, & Stoffer, 2006). Figure 2 shows the biosensor concept and

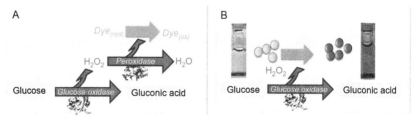

Figure 2 Schematic representation of the working principle of colorimetric enzyme-based assays for the detection of glucose: (A) conventional assay involving the use of HRP and an organic dye. (B) CeO₂ NPs-based assay for the detection of glucose. This principle has been demonstrated on paper surfaces with both the enzyme and the NPs stabilized on paper to create a reagentless enzyme assay (Ornatska et al., 2011). (See the color plate.)

detection mechanism by comparison with a conventional colorimetric assay involving the use of a soluble redox dye and the peroxidase enzyme to quantity the enzymatically generated H_2O_2. In the CeO_2 biosensor, the CeO_2 particles replace both the peroxidase and the corresponding organic dye, significantly increasing the stability and robustness of the assay, reducing the price, while also serving as a robust support for the enzyme.

The method enabled a detection limit of 0.5 mM glucose with a linear range up to 100 mM and a reproducibility of 4.3%. The CeO_2 NPs paper sensor is fully reversible and can be reused for at least 10 consecutive measurement cycles. The only analysis step needed is the addition of the substrate. The biosensor does not require external reagents, as all the sensing components (enzyme- and color-generating reagent—CeO_2) are fixed onto the paper platform. The sensor can be stored for at least 79 days at room temperature while maintaining the same analytical performance over this time.

Robustness, ease-of-use, high reproducibility, and stability of the assay set these sensors apart among other assays that involve the use of sensitive dyes and enzymes. Sensors based on these particles are portable and operate similar to a small sensor patch, allowing for rapid and sensitive detection of target analytes in various environments including food, clinical, and environmental matrices. Thus, replacement of peroxidase and color-generating reagent is no longer required with this strategy. This improvement is significant and can be used as a general platform in the assays of substrates that use oxidases and generate hydrogen peroxide as the product.

4.2 Materials

4.2.1 Sensing Elements

Filter paper (Whatman filter paper #1) used as the paper platform for all sensors was purchased from VWR, Philadelphia, PA (supplier No. 1001-110) and VWR Catalog #28450-106 and the CeO_2 NPs from Sigma Aldrich, St. Louis, MO (Catalog #289744).

4.2.2 CeO$_2$ NP Sensors

CeO_2 NPs used for paper sensor fabrication are of <20 nm in diameter and they are purchased from a variety of suppliers. The example demonstrated here was developed with cerium (IV) oxide NPs (20 wt.% colloidal dispersion in 2.5% acetic acid, 10–20 nm average particle size) from Sigma Aldrich (Catalog #289744). CeO_2 NPs from other suppliers or synthesized in the lab can also be substituted. Different particles may show different color intensities and some may aggregate and may not be uniformly deposited on the paper surface. In general, the particles in aqueous dispersions provided superior characteristics over dry CeO_2 powders for this type of assays (Sharpe et al., 2013).

4.2.3 Enzyme and Other Reagents
GOX (G6641), glucose (G5767), aminopropyltriethoxysilane (APTS), chitosan, glutaraldehyde, H_2O_2, and sodium phosphate were purchased from Sigma Aldrich.

4.3 Methods
4.3.1 Fabrication of CeO_2-Enzyme Paper
The sensing paper is comprised of CeO_2 NPs coimmobilized with GOX on cellulose paper (Ornatska et al., 2011). First CeO_2 modified filter paper was made by immersing 11 cm diameter filter paper rounds into baths of 3% CeO_2 NP dispersion for 10 min and dried for several hours at 75°C. The NP dispersion was sonicated before use. After drying, the CeO_2 NP paper was dip-soaked in 5% APTS in ethanol for 10 min to stabilize the NPs. The paper with stabilized NPs was then dried for 10 min at 100°C. The APTS stabilization procedure enables the formation of siloxane bridges between the OH-rich surface of the cellulosic fibers and the hydroxylated ceria through hydrogen bonding (Salon, Abdelmouleh, Boufi, Belgacem, & Gandini, 2005), effectively attaching the particles to the paper matrix. CeO_2 NPs prepared by this method show good uniformity and surface coverage. The amino-functional groups of the APTS are further used for the grafting of GOX. We note that the CeO_2 NPs maintained their color changing properties in response to H_2O_2 after silanization, which demonstrates that the chemistry used here is compatible with NPs.

4.3.2 Enzyme Immobilization on the CeO_2-Silane Paper
To immobilize the enzyme on the NP-coated, modified cellulose matrix, the silane-CeO_2 NP paper disks were soaked in 1% chitosan solution (in 0.5% succinic acid) for 10 min and air dried for 5–10 min. The biopolymer chitosan coating on the silica paper increases enzyme stability while providing access to the enzyme active site. The paper was then treated with 5% glutaraldehyde for 1 min for crosslinking the enzyme and various components, dipped in water and air dried for 5–10 min. A solution of 20 µL of 9 mg/mL GOX (190.8 U/mL) was applied onto the CeO_2 paper to load the enzyme and modified paper sensors were then rinsed with phosphate buffer at pH 7.4 and air dried.

4.4 Detection and Measurement Procedure

Glucose detection was achieved with the CeO_2–GOX modified paper immersed in glucose solutions of concentrations ranging from 0.5–500 mM in 0.05 M phosphate buffer pH 7.4. Alternatively, samples containing glucose solutions can be added to each CeO_2–GOX testing disk. A standard calibration curve was created with increasing glucose concentrations, displaying the relationship of color intensity to the concentration of glucose. Typical calibration curves are linear but logarithmic relationship can also be obtained under specific conditions. Calibration curves are intended for use in quantitative analysis of samples. Color responses can be read once sensors are fabricated and dried.

The CeO_2 sensor displays an immediate dark-orange color upon the addition of H_2O_2 in the concentration range from 2.5 to 100 mM. When the CeO_2 NPs were coimmobilized with GOX on paper, addition of glucose generates H_2O_2 which reacts with the immobilized NPs, generating the dark-orange color in a concentration-dependent manner. No change in color was observed in the absence of the enzyme, which demonstrates that the color formed is due to the enzymatically generated H_2O_2 and that the sensor is selective for glucose. The color is evenly distributed on the surface of the biosensor indicating uniform distribution and surface coverage of the NPs and enzyme onto the paper. This biosensor can be used to detect glucose at concentrations ranging between 0.5 and 100 mM (Figure 3).

The biosensors are reproducible with a standard deviation below 5%. For example, when 11 biosensors were tested, the average color intensity after addition of 100 mM glucose was 102.2 (±4.4). The biosensors can be stored for at least 79 days at room temperature. Stability was evaluated by periodic testing of the response of GOX biosensors from the same lot (e.g., prepared at the same time) with the addition of 100 mM glucose. The sensors stored in the refrigerator showed no decrease in response after 79 days over this time period. The response of biosensors stored at room temperature decreased by 10% after 50 days and by 15% after 79 days of storage. This also demonstrates effective immobilization of the enzyme on the paper disk and enhancement of enzyme stability in the sensing matrix.

For quantification of the color observed with the glucose solutions, the CeO_2–GOX paper sensors were scanned with a conventional office scanner. Images were analyzed using Adobe Photoshop or the ImageJ software. The colors of the paper were assessed using the eyedropper tool on Photoshop, which took an average of 31 × 31 pixels (or 961 pixels) close to the center of

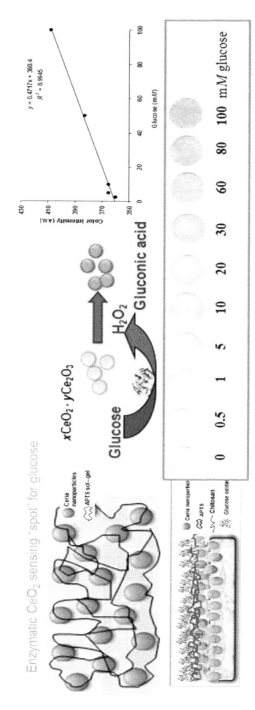

Figure 3 CeO$_2$–GOX paper-based biosensor and colorimetric response to glucose concentrations ranging from 0 to 100 mM glucose. (See the color plate.)

the paper sensors. The blue color intensity was found to be the most sensitive option as blue is the complementary color to yellow/orange. The intensity of the color was used to construct the calibration curve and characterize the performance of the assay. Alternatively, the Pantone CapSure® handheld color analysis device may be used for the creation of a color database due to its ability to generate unique color ID codes and RGB values that can be referenced by all users of the same device.

For analysis of unknown concentrations of the glucose samples, sample was added to the CeO_2–GOX paper and the intensity of the color measured. An example of the application of the as-prepared sensing paper is for the detection of glucose in serum, in the presence of a variety of biological materials which are potential interfering substances. Glucose concentration in serum was determined with these biosensors (3.71 mM), as deduced from the calibration curve using the standard addition method.

5. COMMENTS ON THE METHOD
5.1 NP Choice

The surface characteristics of NPs and their chemical nature dictate the interaction and performance of the enzyme biosensor. The surface functionality can affect enzyme activity and binding characteristics (Sheldon, 2007). Surface reactivity also varies with the synthetic procedure and additives/surface coatings used to stabilize the particles (Sharpe et al., 2013) and can increase by addition of dopants such as Pt (Sardesai et al., 2014). One outstanding question is whether engineered NPs can be incorporated in a biologically relevant environment while conserving their useful properties for sensing. In their vast majority, synthesized CeO_2 NPs contain a surface coating to increase stability and dispersibility. The functional group of the surface ligand is used to attach enzymes and it can vary among particles. In principle, we recommend particles with hydroxyl, amino, or carboxylic terminal groups. The color changing properties and reactivity of these particles can vary greatly from one particle type to another. NPs should be tested for color formation with H_2O_2 before their use in the fabrication of the biosensors. Size, uniformity, and homogeneity of the NPs can affect their optical properties as well as the overall performance of biosensor. CeO_2 NPs should first be tested to verify uniform deposition and stability on paper. Uniform coating of the NPs in the paper can be tested by the addition of H_2O_2 to the CeO_2-modified paper and examining the color distribution.

5.2 Enzyme and NP Immobilization

Both the enzyme and the NP can be printed on the paper surface to enable scalable fabrication of these devices. Natural biopolymers, chitosan, and alginate can serve as printing inks. These polymers are excellent materials for the immobilization and stabilization of enzymes and they form transparent films. A recently developed colorimetric paper-based portable enzyme biosensor with the enzyme attached to deposited layers of chitosan and alginate, provided extended stability and functionality of these sensors over 250 days at room temperature (Alkasir, Rossner, & Andreescu, 2015). Similarly, enzyme–polymer conjugates were interlocked in the fibers of paper for enhanced stability and activity of GOX (Riccardi et al., 2016). We used printing conditions to mass-produce CeO_2–GOX assay and CeO_2 NPs are also synthesized in situ by printing to improve stability and uniformity of the NPs. With suitable optimization of printing ink formulations, the entire sensing assembly, including the enzyme and particle attachment, could be fabricated by inkjet or by 3D printing.

5.3 Testing for Interferences

The effect of interfering compounds should be tested for all sensing applications. Control experiments with paper sensors in the absence of enzyme are recommended to check for potential interferences from the sample, in particular, from compounds that may nonspecifically react or alter the color changing properties of CeO_2 NPs.

Some compounds that have phenolic structures such as dopamine or phenol can generate a color change by forming charge transfer complexes at the CeO_2 NP surface but these changes are only observed if the compounds are present at millimolar concentrations. Interferences can be rejected by the addition of permselective or electrostatic membranes. For example, chitosan has electrostatic rejection properties against ascorbic acid and uric acid (Ozel, Wallace, & Andreescu, 2011) and it can be deposited on the sensing surface to prevent access of these compounds to the active NP platform. Thus, advantages and limitations of these NPs are to be kept in mind in the rational design of biosensors based on CeO_2 NPs.

5.4 Color Analysis

Color analysis of sensor responses can be achieved via smartphone and tablets with the use of an appropriate color analysis application.

ACKNOWLEDGMENTS
This work was funded by the National Science Foundation Grant No. 0954919. Any opinions, findings, and conclusions or recommendations expressed in this material are those of the author(s) and do not necessarily reflect the views of the National Science Foundation or Clarkson University.

REFERENCES
Abe, K., Suzuki, K., & Citterio, D. (2008). Inkjet-printed microfluidic multianalyte chemical sensing paper. *Analytical Chemistry, 80*, 6928–6934.
Ai, H., Jones, S., & Lvov, Y. (2003). Biomedical applications of electrostatic layer-by-layer nano-assembly of polymers, enzymes, and nanoparticles. *Cell Biochemistry and Biophysics, 39*, 23–43.
Alkasir, R. S. J., Rossner, A., & Andreescu, S. (2015). Portable colorimetric paper-based biosensing device for the assessment of bisphenol A in indoor dust. *Environmental Science & Technology, 49*, 9889–9897.
Andreescu, D., Bulbul, G., Ozel, R. E., Hayat, A., Sardesai, N., & Andreescu, S. (2014). Applications and implications of nanoceria reactivity: Measurement tools and environmental impact. *Environmental Science: Nano, 1*(5), 445–458.
Andreescu, S., Njagi, J., & Ispas, C. (2008). Nanostructured materials for enzyme immobilization and biosensors. *New Frontiers of Organic and Composite Nanotechnology*, 355–394.
Ansari, A. A., Solanki, P. R., & Malhotra, B. D. (2009). Hydrogen peroxide sensor based on horseradish peroxidase immobilized nanostructured cerium oxide film. *Journal of Biotechnology, 142*(2), 179–184.
Ardao, I., Comenge, J., Benaiges, M. D., Álvaro, G., & Puntes, V. F. (2012). Rational nanoconjugation improves biocatalytic performance of enzymes: Aldol addition catalyzed by immobilized rhamnulose-1-phosphate aldolase. *Langmuir, 28*(15), 6461–6467.
Arruda, D. L., Wilson, W. C., Nguyen, C., Yao, Q. W., Caiazzo, R. J., Talpasanu, I., et al. (2009). Microelectrical sensors as emerging platforms for protein biomarker detection in point-of-care diagnostics. *Expert Review of Molecular Diagnostics, 9*(7), 749–755.
Asati, A., Santra, S., Kaittanis, C., Nath, S., & Perez, J. M. (2009). Oxidase-like activity of polymer-coated cerium oxide nanoparticles. *Angewandte Chemie, International Edition, 48*(13), 2308–2312.
Babko, A. K., & Volkova, A. I. (1954). The colored peroxide complex of cerium. *Ukrains'kii Khemichnii Zhurnal, 20*, 211–215.
Baron, R., Willner, B., & Willner, I. (2007). Biomolecule-nanoparticle hybrids as functional units for nanobiotechnology. *Chemical Communications, 4*, 323–332.
Czeslik, C., & Winter, R. (2001). Effect of temperature on the conformation of lysozyme adsorbed to silica particles. *Physical Chemistry Chemical Physics, 3*(2), 235–239.
Das, M., Patil, S., Bhargava, N., Kang, J.-F., Riedel, L. M., Seal, S., et al. (2007). Autocatalytic ceria nanoparticles offer neuroprotection to adult rat spinal cord neurons. *Biomaterials, 28*(10), 1918–1925.
Datta, S., Christena, L. R., & Rajaram, Y. R. S. (2013). Enzyme immobilization: An overview on techniques and support materials. *3 Biotech, 3*(1), 1–9.
Davis, V. T., & Thompson, J. S. (2002). Measurement of the electron affinity of cerium. *Physical Review Letters, 88*(7), 073003.
Demin, S., & Hall, E. A. H. (2009). Breaking the barrier to fast electron transfer. *Bioelectrochemistry, 76*(1–2), 19–27.
Ding, S., Cargill, A. A., Medintz, I. L., & Claussen, J. C. (2015). Increasing the activity of immobilized enzymes with nanoparticle conjugation. *Current Opinion in Biotechnology, 34*, 242–250.

Ding, X., Srinivasan, B., & Tung, S. (2015). Development and applications of portable biosensors. *Journal of Laboratory Automation, 20*(4), 365–389.

Dungchai, W., Chailapakul, O., & Henry, C. S. (2009). Electrochemical detection for paper-based microfluidics. *Analytical Chemistry, 81*(14), 5821–5826.

Dungchai, W., Chailapakul, O., & Henry, C. S. (2010). Use of multiple colorimetric indicators for paper-based microfluidic devices. *Analytica Chimica Acta, 674*(2), 227–233.

Dutta, P., Pal, S., Seehra, M. S., Shi, Y., Eyring, E. M., & Ernst, R. D. (2006). Concentration of Ce^{3+} and oxygen vacancies in cerium oxide nanoparticles. *Chemistry of Materials, 18*(21), 5144–5146.

Elghanian, R., Storhoff, J. J., Mucic, R. C., Letsinger, R. L., & Mirkin, C. A. (1997). Selective colorimetric detection of polynucleotides based on the distance-dependent optical properties of gold nanoparticles. *Science, 277*(5329), 1078–1081.

Gabig-Ciminska, M. (2006). Developing nucleic acid-based electrical detection systems. *Microbial Cell Factories, 5*, 9.

Gowers, S. A., Curto, V. F., Seneci, C. A., Wang, C., Anastasova, S., Vadgama, P., et al. (2015). 3D printed microfluidic device with integrated biosensors for online analysis of subcutaneous human microdialysate. *Analytical Chemistry, 87*(15), 7763–7770.

Heller, A. (1990). Electrical wiring of redox enzymes. *Accounts of Chemical Research, 23*(5), 128–134.

Homaei, A. A., Sariri, R., Vianello, F., & Stevanato, R. (2013). Enzyme immobilization: An update. *Journal of Chemical Biology, 6*(4), 185–205.

Hossain, S. M. Z., Luckham, R. E., Smith, A. M., Lebert, J. M., Davies, L. M., Pelton, R. H., et al. (2009). Development of a bioactive paper sensor for detection of neurotoxins using piezoelectric inkjet printing of sol–gel-derived bioinks. *Analytical Chemistry, 81*(13), 5474–5483.

Ispas, C. R., Crivat, G., & Andreescu, S. (2012). Review: Recent developments in enzyme-based biosensors for biomedical analysis. *Analytical Letters, 45*(2–3), 168–186.

Jia, H., Zhu, G., & Wang, P. (2003). Catalytic behaviors of enzymes attached to nanoparticles: The effect of particle mobility. *Biotechnology and Bioengineering, 84*(4), 406–414.

Jo, D. H., Kim, J. H., Lee, T. G., & Kim, J. H. (2015). Size, surface charge, and shape determine therapeutic effects of nanoparticles on brain and retinal diseases. *Nanomedicine: Nanotechnology, Biology and Medicine, 11*(7), 1603–1611.

Khan, R., & Dhayal, M. (2008). Electrochemical studies of novel chitosan/TiO_2 bioactive electrode for biosensing application. *Electrochemistry Communications, 10*(2), 263–267.

Kim, J., Grate, J. W., & Wang, P. (2006). Nanostructures for enzyme stabilization. *Chemical Engineering Science, 61*(3), 1017–1026.

Lata, J. P., Gao, L., Mukai, C., Cohen, R., Nelson, J. L., Anguish, L., et al. (2015). Effects of nanoparticle size on multilayer formation and kinetics of tethered enzymes. *Bioconjugate Chemistry, 26*(9), 1931–1938.

Li, Y. F., Liu, Z. M., Liu, Y. L., Yang, Y. H., Shen, G. L., & Yu, R. Q. (2006). A mediator-free phenol biosensor based on immobilizing tyrosinase to ZnO nanoparticles. *Analytical Biochemistry, 349*(1), 33–40.

Liu, J. W., Mazumdar, D., & Lu, Y. (2006). A simple and sensitive "dipstick" test in serum based on lateral flow separation of aptamer-linked nanostructures. *Angewandte Chemie, International Edition, 45*(47), 7955–7959.

Mabey, D., Peeling, R. W., Ustianowski, A., & Perkins, M. D. (2004). Diagnostics for the developing world. *Nature Reviews. Microbiology, 2*(3), 231–240.

Martinez, A., Phillips, S., Butte, M., & Whitesides, G. (2007). Patterned paper as a platform for inexpensive, low-volume, portable bioassays. *Angewandte Chemie, International Edition, 46*(8), 1318–1320.

Martinez, A. W., Phillips, S. T., Whitesides, G. M., & Carrilho, E. (2010). Diagnostics for the developing world: Microfluidic paper-based analytical devices. *Analytical Chemistry*, *82*(1), 3–10.

Metters, J. P., Houssein, S. M., Kampouris, D. K., & Banks, C. E. (2013). Paper-based electroanalytical sensing platforms. *Analytical Methods UK*, *5*(1), 103–110.

Naruse, J., Hoa, L. Q., Sugano, Y., Ikeuchi, T., Yoshikawa, H., Saito, M., et al. (2011). Development of biofuel cells based on gold nanoparticle decorated multi-walled carbon nanotubes. *Biosensors and Bioelectronics*, *30*(1), 204–210.

Nie, Z. H., Nijhuis, C. A., Gong, J. L., Chen, X., Kumachev, A., Martinez, A. W., et al. (2010). Electrochemical sensing in paper-based microfluidic devices. *Lab on a Chip*, *10*(4), 477–483.

Njagi, J., Chernov, M. M., Leiter, J. C., & Andreescu, S. (2010). Amperometric detection of dopamine in vivo with an enzyme based carbon fiber microbiosensor. *Analytical Chemistry*, *82*(3), 989–996.

Njagi, J., Ispas, C., & Andreescu, S. (2008). Mixed ceria-based metal oxides biosensor for operation in oxygen restrictive environments. *Analytical Chemistry*, *80*(19), 7266–7274.

Ornatska, M., Sharpe, E., Andreescu, D., & Andreescu, S. (2011). Paper bioassay based on ceria nanoparticles as colorimetric probes. *Analytical Chemistry*, *83*(11), 4273–4280.

Ozel, R. E., Ispas, C., Ganesana, M., Leiter, J. C., & Andreescu, S. (2014). Glutamate oxidase biosensor based on mixed ceria and titania nanoparticles for the detection of glutamate in hypoxic environments. *Biosensors & Bioelectronics*, *52*, 397–402.

Ozel, R. E., Wallace, K. N., & Andreescu, S. (2011). Chitosan coated carbon fiber microelectrode for selective in vivo detection of neurotransmitters in live zebrafish embryos. *Analytica Chimica Acta*, *695*(1–2), 89–95.

Pelton, R. (2009). Bioactive paper provides a low-cost platform for diagnostics. *TrAC Trends in Analytical Chemistry*, *28*(8), 925–942.

Pingarrón, J. M., Yáñez-Sedeño, P., & González-Cortés, A. (2008). Gold nanoparticle-based electrochemical biosensors. *Electrochimica Acta*, *53*(19), 5848–5866.

Pirmohamed, T., Dowding, J. M., Singh, S., Wasserman, B., Heckert, E., Karakoti, A. S., et al. (2010). Nanoceria exhibit redox state-dependent catalase mimetic activity. *Chemical Communications*, *46*(16), 2736–2738.

Popat, A., Hartono, S. B., Stahr, F., Liu, J., Qiao, S. Z., & Qing Lu, G. (2011). Mesoporous silica nanoparticles for bioadsorption, enzyme immobilisation, and delivery carriers. *Nanoscale*, *3*(7), 2801–2818.

Preda, G., Migani, A., Neyman, K. M., Bromley, S. T., Illas, F., & Pacchioni, G. (2011). Formation of superoxide anions on ceria nanoparticles by interaction of molecular oxygen with Ce^{3+} sites. *The Journal of Physical Chemistry C*, *115*(13), 5817–5822.

Radhakumary, C. S. K. (2011). Naked eye detection of glucose in urine using glucose oxidase immobilized gold nanoparticle. *Analytical Chemistry*, *83*, 2829–2833. http://dx.doi.org/10.1021/ac1032879.

Reches, M., Mirica, K. A., Dasgupta, R., Dickey, M. D., Butte, M. J., & Whitesides, G. M. (2010). Thread as a matrix for biomedical assays. *ACS Applied Materials & Interfaces*, *2*(6), 1722–1728.

Riccardi, C. M., Mistri, D., Hart, O., Anuganti, M., Lin, Y., Kasi, R. M., et al. (2016). Covalent interlocking of glucose oxidase and peroxidase in the voids of paper: Enzyme-polymer "spider-webs" *Chemical Communications*, *52*, 2593–2596.

Salon, M. C. B., Abdelmouleh, M., Boufi, S., Belgacem, M. N., & Gandini, A. (2005). Silane adsorption onto cellulose fibers: Hydrolysis and condensation reactions. *Journal of Colloid and Interface Science*, *289*(1), 249–261.

Sardesai, N. P., Ganesana, M., Karimi, A., Leiter, J. C., & Andreescu, S. (2015). Platinum-doped ceria based biosensor for in vitro and in vivo monitoring of lactate during hypoxia. *Analytical Chemistry*, *87*(5), 2996–3003.

Sardesai, N. P., Karimi, A., & Andreescu, S. (2014). Engineered Pt-doped nanoceria for oxidase-based bioelectrodes operating in oxygen-deficient environments. *ChemElectroChem, 1*(12), 2082–2088.

Schoemaker, H. E., Mink, D., & Wubbolts, M. G. (2003). Dispelling the myths— Biocatalysis in industrial synthesis. *Science, 299*(5613), 1694–1697.

Sharpe, E., Bradley, R., Frasco, T., Jayathilaka, D., Marsh, A., & Andreescu, S. (2014). Metal oxide based multisensor array and portable database for field analysis of antioxidants. *Sensors and Actuators B: Chemical, 193*, 552–562.

Sharpe, E., Frasco, T., Andreescu, D., & Andreescu, S. (2013). Portable ceria nanoparticle-based assay for rapid detection of food antioxidants (NanoCerac). *Analyst, 138*(1), 249–262.

Sheldon, R. A. (2007). Cross-linked enzyme aggregates (CLEA®s): Stable and recyclable biocatalysts. *Biochemical Society Transactions, 35*(6), 1583–1587.

Topoglidis, E., Cass, A. E. G., O'Regan, B., & Durrant, J. R. (2001). Immobilisation and bioelectrochemistry of proteins on nanoporous TiO_2 and ZnO films. *Journal of Electroanalytical Chemistry, 517*(1–2), 20–27.

Wang, D., Kang, Y., Doan-Nguyen, V., Chen, J., Küngas, R., Wieder, N. L., et al. (2011). Synthesis and oxygen storage capacity of two-dimensional ceria nanocrystals. *Angewandte Chemie, International Edition, 50*(19), 4378–4381.

Wilson, G. S., & Hu, Y. (2000). Enzyme-based biosensors for in vivo measurements. *Chemical Reviews, 100*(7), 2693–2704.

Won, Y.-H., Aboagye, D., Jang, H. S., Jitianu, A., & Stanciu, L. A. (2010). Core/shell nanoparticles as hybrid platforms for the fabrication of a hydrogen peroxide biosensor. *Journal of Materials Chemistry, 20*(24), 5030–5034.

Xu, J., Harmer, J., Li, G., Chapman, T., Collier, P., Longworth, S., et al. (2010). Size dependent oxygen buffering capacity of ceria nanocrystals. *Chemical Communications, 46*(11), 1887–1889.

Yu, P., Hayes, S. A., O'Keefe, T. J., O'Keefe, M. J., & Stoffer, J. O. (2006). The phase stability of cerium species in aqueous systems—II. The $Ce(III/IV)-H_2O-H_2O_2/O_2$ systems. Equilibrium considerations and Pourbaix diagram calculations. *Journal of the Electrochemical Society, 153*(1), C74–C79.

Zhang, F., Wang, P., Koberstein, J., Khalid, S., & Chan, S. W. (2004). Cerium oxidation state in ceria nanoparticles studied with X-ray photoelectron spectroscopy and absorption near edge spectroscopy. *Surface Science, 563*(1–3), 74–82.

Zhang, M., Wang, H., Wang, X., & Li, W. (2006). Complex impedance study on nano-CeO_2 coating TiO_2. *Materials & Design, 27*(6), 489–493.

Zhao, W. A., Ali, M. M., Aguirre, S. D., Brook, M. A., & Li, Y. F. (2008). Paper-based bioassays using gold nanoparticle colorimetric probes. *Analytical Chemistry, 80*(22), 8431–8437.

Zhao, W., Brook, M. A., & Li, Y. F. (2008). Design of gold nanoparticle-based colorimetric biosensing assays. *Chembiochem, 9*(15), 2363–2371.

CHAPTER TEN

Rational Design of Nanoparticle Platforms for "Cutting-the-Fat": Covalent Immobilization of Lipase, Glycerol Kinase, and Glycerol-3-Phosphate Oxidase on Metal Nanoparticles

V. Aggarwal, C.S. Pundir[1]

Department of Biochemistry, Maharshi Dayanand University, Rohtak, Haryana, India
[1]Corresponding author: e-mail address: pundircs@rediffmail.com

Contents

1. Introduction	198
2. Use of Rationally Designed Nanoscaffolds for Enzyme Binding	200
3. Use of Chitosan for Enhancing Nanoparticle Surface Chemistry	201
4. Experimental	202
4.1 Combined Assay of Mixture of Free/Native Lipase, GK and GPO	202
4.2 Combined Assay of Enzymes	202
4.3 Kinetic Properties of Co-Immobilized Lipase, GK, and GPO	215
4.4 Application of Co-Immobilized Lipase, GK, and GPO onto Nanomaterials	219
References	222

Abstract

The aggregates of nanoparticles (NPs) are considered better supports for the immobilization of enzymes, as these promote enzyme kinetics, due to their unusual but favorable properties such as larger surface area to volume ratio, high catalytic efficiency of certain immobilized enzymes, non-toxicity of some of the nanoparticle matrices, high stability, strong adsorption of the enzyme of interest by a number of different approaches, and faster electron transportability. Co-immobilization of multiple enzymes required for a multistep reaction cascade on a single support is more efficient than separately immobilizing the corresponding enzymes and mixing them physically, since products of one enzyme could serve as reactants for another. These products can diffuse much more easily between enzymes on the same particle than diffusion from one particle to the next, in the reaction medium. Thus, co-immobilization of enzymes onto NP aggregates is expected to produce faster kinetics than their individual immobilizations on separate matrices. Lipase, glycerol kinase, and glycerol-3-phosphate oxidase are required for lipid analysis in a cascade reaction, and we describe the co-immobilization of these three enzymes on nanocomposites of zinc oxide nanoparticles (ZnONPs)—chitosan (CHIT) and gold nanoparticles-polypyrrole-polyindole carboxylic acid (AuPPy-Pin5COOH) which are electrodeposited on Pt and Au electrodes, respectively. The kinetic properties and analytes used for amperometric determination of TG are fully described for others to practice in a trained laboratory. Cyclic voltammetry, scanning electron microscopy, Fourier transform infra-red spectra, and electrochemical impedance spectra confirmed their covalent co-immobilization onto electrode surfaces through glutaraldehyde coupling on CHIT-ZnONPs and amide bonding on AuPPy/Pin5COOH. The combined activities of co-immobilized enzymes was tested amperometrically, and these composite nanobiocatalysts showed optimum activity within 4–5 s, at pH 6.5–7.5 and 35 °C, when polarized at a potential between 0.1 and 0.4 V. Co-immobilized enzymes showed excellent linearity within 50–700 mg/dl of the lipid with detection limit of 20 mg/dl for triolein. The half life of co-immobilized enzymes was 7 months, when stored dry at 4 °C which is very convenient for practical applications. Co-immobilized biocatalysts measured triglycerides in the sera of apparently healthy persons and persons suffering from hypertriglyceridemia, which is recognized as a leading cause for heart disease. The measurement of serum TG by co-immobilized enzymes was unaffected by the presence of a number of serum substances, tested as potential interferences. Thus, co-immobilization of enzymes onto aggregates of NPs resulted in improved performance for TG analysis.

1. INTRODUCTION

Triglycerides (TGs) are esters of glycerol and fatty acids in a stoichiometric molar ratio of 1:3, as the very name implies. These are the major components of very low-density lipoproteins (VLDL) and chylomicrons, which are aggregates of TG, phospholipids, proteins, and cholesterol. These are two of the five major classes of lipoproteins that play a very important role in the metabolism as energy sources as well as transporters of dietary fat from the intestines to other parts of the body. Increased level of TGs in the blood stream (>190 mg/dl) leads to atherosclerosis which increases the risk of heart disease, coronary heart disease (CAD), and hypolipoprotenimia. Normally, serum TG level in healthy persons is in the range of 150–190 mg/dl. The increased level of TG in the range 200–500 mg/dl is an alarming situation, while the level >500 mg/dl is troublesome (Nelson, Cox, & Lehninger, 2000). Hence, reliable measurement of TG levels in serum is very important for public health and for diagnosis of heart disease. Among various methods available for TG determination, enzymatic colorimetric method is the most popular. The method involves a cascade of three enzymic reactions followed by a color reaction where the end point provides quantitation of TG. The three major reactions of the assay are catalyzed by three different enzymes, lipase, glycerol kinase (GK), and glycerol-3-phosphate oxidase (GPO), which are shown below:

$$\text{Triglyceride} + 3H_2O \xrightarrow{\text{Lipase}} 3\,\text{fatty acid} + \text{glycerol}$$

$$\text{Glycerol} + \text{ATP} \xrightarrow[Mg^{2+}]{\text{GK}} \text{glycerol-3-phosphate} + \text{ADP}$$

$$\text{Glycerol-3-phosphate} \xrightarrow{\text{GPO}} \text{dihydroxyacetone phosphate} + H_2O_2$$

The color reaction is as following:

$$H_2O_2 + \text{DHBS} + \text{4-aminophenazone} \xrightarrow{\text{Peroxidase}} \text{quinoneimine (pink color)}$$

DHBS, 3,5 dichloro-2-hydroxybenzene sulfonic acid.

However, this enzymatic colorimetric method employing free/native enzymes is expensive, when employed for a large number of samples, due to the high cost of the enzymes needed, and they cannot be recycled readily. These enzymes have poor stability under the experimental conditions encountered in an ordinary laboratory and need to be purchased periodically. Attachment of enzymes onto an insoluble support not only provides their reusability but also realizes the extra stabilization rendered by the multipoint covalent attachment or multisubunit binding of the enzymes on solid supports. In some cases, we and others found that enzymes bound to nanosolids improve enzyme activity, specificity, and/or stability (Barbosa et al., 2015; Kazenwadel, Franzreb, & Rapp, 2015; Kumar & Chaudhari, 2000, 2002). In recent years, solid-bound enzymes have revolutionized the prospects of enzyme applications in industry, and rational design of the solid support has gained major attention. Earlier, the biocatalyst preparation methodology was more like hit-and-trial but a rational design of the surface of the nanoparticles and rational choice of the linkage chemistry have improved the success of the approach, drastically. The rational approach is made possible through intense activity in understanding how enzymes and nanosolids interact at the molecular level, and a number of research groups around the world have contributed to this important activity (Chowdhury, Stromer, Pokharel, & Kumar, 2012; Duff & Kumar, 2009; Mudhivarthi, Bhambhani, & Kumar, 2007; Novak et al., 2016). Enzyme immobilization can be defined as the confinement of enzyme molecules onto/within a support/matrix physically or chemically or both, in such a way that it retains its full activity or most of its activity.

When a series of reactions are to be conducted, as in the case of TG analysis reaction cascade, co-binding of multiple enzymes on the same particle has significant kinetic advantages. In a cascade reaction, the product of one enzyme is the substrate for another, and when all required enzymes are on the same particle, then this intermediary product does not have to diffuse from one particle to another particle to find its corresponding enzyme in the cascade reaction. Therefore, the diffusion distance and time are considerably shortened and such a system is more efficient than a mixture of separately bound enzyme particles, where the intermediary products will have to diffuse from one particle to the next for the continuation of the reaction steps (Minakshi & Pundir, 2008a, 2008b). However, one disadvantage of such a communal enzyme-particle approach is that when one or more enzymes in the assembly are deactivated, the entire assembly is effected.

Of the different techniques available for immobilization such as adsorption, encapsulation, microencapsulation, cross-linking, metal chelation, ionic binding, covalent binding is preferred, as it avoids enzyme leaching, provides comparative stability, durability, reusability, and shelf life. Lipase, GK, and GPO have been co-immobilized onto various supports such as collagen (Winartasaputra, Kutan, & Cuilbault, 1982), alkyl amine glass beads (Kalia & Pundir, 2002), aryl amine glass beads (Kalia & Pundir, 2004) through covalent coupling, and to cellulose acetate (Minakshi & Pundir, 2008a, 2008b), polyvinylchloride (PVC) (Narang, Bhambi, Minakshi, & Pundir, 2010), and polyvinyl alcohol (PVA) (Pundir, Singh, Bharvi, & Narang, 2010) through adsorption. In contrast, intercalation of enzymes in the galleries of nanosheets was enhanced by tuning enzyme charge, and enzyme stability increased at an unprecedented 150-fold (Novak et al., 2016). Such rational design of the nanomaterials is of intense interest for making progress in the cascade reaction catalysis. In contrast, physical adsorption of the enzymes onto membranes has the risk of their leakage during the washing for reuse.

2. USE OF RATIONALLY DESIGNED NANOSCAFFOLDS FOR ENZYME BINDING

Materials when reduced down to 1–100 nm in one or more of their dimensions, exhibit drastic changes in their physical, chemical, optical, magnetic, or electrochemical properties, which lead to exciting applications in bioscience, medicine, environmental science, and various industries. Different nanostructures have been synthesized such as nanoparticles (NPs), nanorods, nanotubes, nanofibers, nanospheres and nanofilms. Of these nanostructures, NPs and clusters of nanoparticles are most often employed for enzyme binding due to the possibility for rational control of the enzyme–solid interactions at these surfaces.

NPs can be organic, inorganic, or biological in nature, and this chapter focuses on zinc oxide nanoparticles (ZnONPs) which are potentially good nanomaterials for enzyme binding because of favorable properties such as their large surface area due to large volume to weight ratio, high catalytic efficiency, non-toxicity, chemical stability, strong adsorption ability, and high isoelectric point, pH 9.5. Nanoporous nature of the ZnONPs also enhances the active surface area for strong adsorption of biomolecules (Singh et al., 2007; Wang et al., 2006; Wei et al., 2006).

3. USE OF CHITOSAN FOR ENHANCING NANOPARTICLE SURFACE CHEMISTRY

Chitosan (CHIT) is an interesting biopolymer for the binding of a variety of biomolecules (including enzymes), because of excellent film-forming ability, high permeability, good mechanical strength, non-toxicity, biocompatibility, low cost, and easy availability. Moreover, chemical modification of the amino groups of CHIT provides hydrophilic environment, which may be necessary for the retention of the native-like structures of most biomolecules (Jiang et al., 2007; Liao, Lin, & Wu, 2005).

Chemical structure of chitosan *Source: Wikipedia.*

Likewise, gold polypyrrole (AuPPy) nanocomposites have a remarkable property. They adhere easily to Au electrodes and show a better electrocatalytic reduction than bare Au electrode. Further, AuNPs AuPPy nanocomposites help in faster electron transfer to the electrodes. Poly (indole-5-carboxylic acid) (Pin5COOH) not only facilitates the electron transfer between enzymes and electrode but also provides free COOH groups for the covalent binding with –NH_2 groups on the surface of enzymes.

This chapter describes the co-immobilization of lipase, GK, and GPO onto nanocomposites of CHIT-ZnONPs electrodeposited that are onto Pt and AuPPy–Pin5COOH electrodeposited onto Au electrodes. The kinetic properties and applications of these electrodes in the construction of amperometric TG biosensors are examined and improved analytic performance has been noted (Narang & Pundir, 2011; Narang, Chauhan, Rani, & Pundir, 2012).

4. EXPERIMENTAL

4.1 Combined Assay of Mixture of Free/Native Lipase, GK and GPO

4.1.1 Preparation of Mixture of Enzymes

Materials: Lipase (from porcine pancreas), glycerol kinase (GK, from *Cellulomonas*), glycerol-3-phosphate oxidase (GPO, from *Aerococcus viridans*), $MgCl_2$, ATP, potassium ferrocyanide, 4-aminophenazone, 3,5 dichloro-2-hydroxybenzene sulfonic acid (DHBS), triolein solution, and Triton X-100.

Equipment: UV–vis spectrophotometer.

Dissolve 2.0 mg of lipase (from porcine pancreas; 40–70 U/mg) + 1.0 mg GK (from *Cellulomonas* sp; 25–75 U/mg) + 0.1 mg GPO (from *Aerococcus viridans*; 113 U/mg) into 1.0 ml of 0.02 M sodium phosphate buffer (PB), pH 7.0, of 50–100 units of each of the three enzymes in the mixture and store at 4 °C until use.

4.2 Combined Assay of Enzymes

The combined assay of mixture of lipase, GK and GPO in solution is carried out in a 15-ml conical flask wrapped with a black paper, as described previously (Fossati & Prencipe, 1982). First, prepare a reaction mixture consisting of 0.54 µmol $MgCl_2$, 0.63 µmol ATP, 1.0 µmol potassium ferrocyanide, 0.27 µmol 4-aminophenazone, 1.0 µmol DHBS, and 1.0 µmol of Triton X-100 in a total volume of 1 l of 0.1 M PB, pH 7.0. To 900 µl of this reaction mixture, add 50 µl of enzyme mixture and preincubate at 37 °C for 5 min. To start the reaction, add 50 µl of 22.6 mM triolein solution (commercially available at a concentration of 200 mg/dl in the kit for enzymic colorimetric determination for TGs supplied by Span Diagnostics, Surat, Gujarat, India) to this mixture and incubate it at 37 °C for 15 min. Read absorbance at 510 nm (A_{510}) of the reaction mixture against the control/blank in Spectronic 20D$^+$ (Thermo Scientific, USA) and calculate µmol of H_2O_2 generated, using a standard curve generated with known concentrations of H_2O_2. The unit activity of mixture of enzyme is expressed as µmol H_2O_2 generated in the assay/min/ml of mixture of enzymes.

4.2.1 Covalent Co-Immobilization of Lipase, GK, and GPO onto Nanocomposite of ZnONPs/CHIT Electrodeposited onto Pt Electrode

4.2.1.1 Preparation of ZnONPs
Materials: Zinc nitrate ($Zn(NO_3)_2 \cdot 4H_2O$), NaOH, and ethanol.
Equipment: Magnetic stirrer.

To synthesize ZnONPs, prepare 0.9 M sodium hydroxide (NaOH) solution and 0.45 M zinc nitrate ($Zn(NO_3)_2 \cdot 4H_2O$) in double distilled water (DW). Heat NaOH solution (100 ml) at 55 °C and then add 50 ml $Zn(NO_3)_2$ solution dropwise (slowly for 40 min) into hot NaOH solution, under high-speed stirring on a hot plate (Remi, New Delhi, India). Seal the beaker during this process for 2 h, clean the precipitated ZnONPs with DW, until the pH of wash is 7.0, followed by washing once with 10 ml ethanol, and then air dry in hot air oven (NSW, New Delhi) at 60 °C (Lata, Batra, Singala, & Pundir, 2013).

4.2.1.2 Characterization of ZnONPs
Techniques: Transmission electron microscopy and powder XRD.

Before characterization of NPs, sample preparation is required. In case of transmission electron microscopy (TEM), prepare NP suspension in ethanol by sonication in Misonix ultrasonic liquid processor (XL 2000 series, Newtown City, USA) for 15 min and for X-ray diffraction (XRD), grind NPs powder in pestle and mortar until it becomes a fine powder and press it into sample holder of X-ray diffractometer (122 Rigaku, D/Max2550, Tokyo, Japan) to get a smooth plane surface for XRD study. Two techniques are employed to characterize the following properties of ZnONPs.

4.2.1.3 Shape and Size
TEM images of ZnONPs taken in transmission electron microscope (JEOL 2100 F, USA) showed their spherical structure with average size of 50 nm but their aggregates had a flower-like structure (Fig. 1) and XRD pattern shows peaks which are consistent with the peaks of wurtzite ZnO (JCPDS card No. 36-1451, $a=3.249$ Å, $c=5.206$ Å) with high crystallinity (Fig. 2). Furthermore, no other characteristic peaks are observed, revealing no impurities in the samples (Narang & Pundir, 2011).

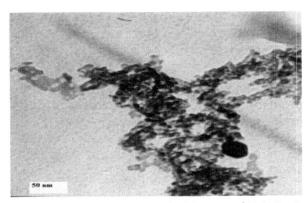

Figure 1 Transmission electron microscopic (TEM) image of ZnONPs. *Reprinted from Narang and Pundir (2011), with permission from Elsevier.*

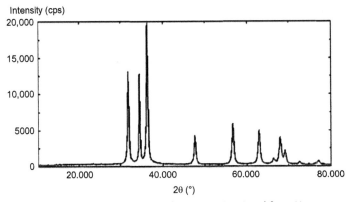

Figure 2 X-ray diffraction (XRD) pattern of ZnONPs. *Reprinted from Narang and Pundir (2011), with permission from Elsevier.*

4.2.1.4 Preparation of ZnONPs-CHIT Composite

Equipment: Ultrasonicator, scanning electron microscope, and FTIR spectrophotometer.

Disperse ZnONP powder (5 mg) into transparent CHIT solution (0.5 g of CHIT/10 ml of 0.05 M acetate buffer solution, pH 5.0) and keep it overnight at room temperature, followed by magnetic stirring for 30 min to mix the contents, and then sonicate it for about 1 h at room temperature. The hybrid nanocomposite of ZnO NP-CHIT is formed overnight, which is confirmed by comparing its scanning electron microscopic (SEM) images, taken in scanning electron microscope (Joel JSM 6510, Japan) and Fourier transform infra-red (FTIR) spectra taken in FTIR spectrophotometer (iS10,

Thermo Scientific, USA) and compared with the corresponding data obtained from CHIT control samples (Narang & Pundir, 2011).

4.2.1.5 Deposition of ZnO NPs-CHIT Composite onto Pt Electrode

Deposit nanocomposite film onto Pt electrode (PtE) by dipping the electrode into solution of hybrid nanocomposite of ZnONPs–CHIT overnight. Activate hybrid nanocomposite film by dipping the modified electrode into 2.5% glutaraldehyde in 0.02 M PB, pH 7.0, and keep it for 2 h at room temperature. Remove this activated electrode from buffer and wash it thoroughly with 0.02 M PB, pH 7.0, until the pH of wash is neutral and apply it onto Pt electrode (surface area of 0.25 cm^2) (Narang & Pundir, 2011).

4.2.1.6 Co-Immobilization of Lipase, GK, and GPO onto ZnO NPs-CHIT/PtE

Dip the nanocomposite film modified Pt electrode into 1 ml mixture of lipase, GK, and GPO dissolved in 0.02 M PB, pH 7.0, and keep it overnight at 4 °C for the covalent immobilization. Wash the enzyme electrode with DW for number of times, followed by 50 mM PB, pH 7.0, to remove unbound enzyme, until the pH of wash is 7.0. Keep the enzyme electrode, i.e., enzymes/nanoZnO-CHIT/Pt electrode undisturbed for about 12 h at 4 °C (Narang & Pundir, 2011).

4.2.1.7 Chemistry of Immobilization of Enzymes on ZnONPs/CHIT Nanocomposite

The covalent co-immobilization of enzymes onto ZnONPs–CHIT composite involves the activation of CHIT of ZnONPs–CHIT nanocomposite by glutaraldehyde (a bifunctional reagent, a powerful cross-linker, and a versatile tool in covalent immobilization of enzymes) and then interaction between activated CHIT and enzymes through Schiff base as following:

$$\underset{\text{Chitosan}}{CHIT-NH_2} + \underset{\text{Glutaraldehyde}}{OHC-(CH_2)_3-CHO} \rightarrow CHIT-NH=CH-(CH_2)-CHO + H_2O$$

$$CHIT-NH=CH(CH_2)_3-CHO + \underset{\text{Enzyme}}{NH_2-E}$$

$$\rightarrow \underset{\text{Immobilized enzymes}}{CHIT-NH=CH-(CH_2)_3-CH=NH-E} + H_2O$$

The steps involved in the binding of enzymes onto nanocomposite of CHIT-ZnONPs are shown in Fig. 3.

Tip: The Schiff base thus formed is prone to hydrolysis unless reduced with a reagent such as sodium borohydride to form the corresponding amine.

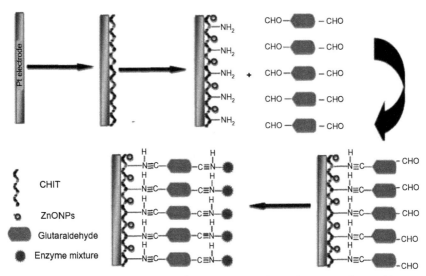

Figure 3 Schematic representation of ZnONPs/CHIT electrode preparation and immobilization of enzymes on it. Step 1+Step 2: Uniform spreading of chitosan solution containing ZnONPs onto Pt electrode; Step 3: Activation of CHIT/ZnONPs/PtE; Step 4: Covalent co-immobilization of mixture of lipase, GK, and GPO onto activated electrode. *Reprinted from Narang and Pundir (2011), with permission from Elsevier.* (See the color plate.)

4.2.2 Covalent Co-Immobilization of Lipase, GK, and GPO onto Nanocomposite Film of Pin5COOH/AuPPy Electrodeposited onto Au Electrode

4.2.2.1 Construction of Gold Polypyrrole (AuPPy) Nanocomposite

Materials: Pyrrole, $HAuCl_4$, and cetyl trimethyl ammonium bromide (CTAB).

A nanocomposite of AuPPy is prepared by direct redox reaction between pyrrole monomers and $HAuCl_4$ as described (Miao, Wu, Chen, Liu, & Qiu, 2008). Mix 10 μl pyrrole into 2 ml of 0.5 M H_2SO_4 containing 10 mM CTAB and then add 10 μl of 2 mM $HAuCl_4$ aqueous solution dropwise and stir overnight on a magnetic stirrer (Remi, New Delhi). The appearance of transparent light yellow color confirms the formation of stably dispersed AuPPy colloid (Narang et al., 2012).

4.2.2.2 TEM Characterization of AuPPy Nanocomposite

TEM is employed to confirm the formation of AuPPy nanoparticles. The TEM image of this nanocomposite shows spherical shape with diameter of 20 nm. In addition to these, small submicron particles are dispersed on the surface; as a result, large agglomerates are seen in the image, which confirm the formation of nanocomposite of AuPPy (Fig. 4) (Narang et al., 2012).

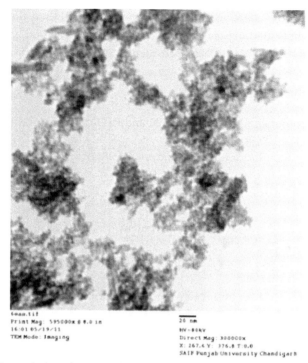

Figure 4 Transmission electron microscopic (TEM) image of AuPPy. *Reprinted from Narang et al. (2012), with permission from Elsevier.*

4.2.2.3 Preparation of Nanocomposites of AuPPy and Pin5COOH

The nanocomposite of AuPPy and Pin5COOH is prepared as described, (Vu et al., 2005) with some modification. To prepare it, add $FeCl_3$ (4.5 g) to a mixture of 1.0 ml AuPPy suspension and 1.0 ml indole monomers (5.0 mg in 20.0 ml DW) under continuous stirring for 2 h, wash the particles of nanocomposite with DW, filter, extract them (in ethanol) for 10 h and then dry at 50 °C.

4.2.2.4 Electrodeposition of Pin5COOH/AuPPynano-Composite onto Au Electrode

Technique: Cyclic voltammetry.

Electropolymerization/deposition of nanocomposite of Pin5COOH/AuPPy onto Au electrode from acetonitrile solution is carried out as follow: Polish an Au (1.5 cm × 0.05 cm) electrode with alumina powder followed by cyclic voltammetry in the potential range, from 0.0 to 0.6 V versus Ag/AgCl, at a scan rate 20 mV/s. Immerse the cleaned Au electrode in

25 ml acetonitrile containing 0.1 g nanocomposite and then apply 10 polymerization cycles in the potential range from 0.0 to 0.6 V and back to 0.0 V at a scan rate of 20 mV/s. After electrodeposition/polymerization, wash the electrode carefully with acetonitrile and finally with DW (Barlett, Dawnson, & Farrington, 1992).

4.2.2.5 Co-Immobilization of Lipase, GK, and GPO onto Pin5COOH/AuPPy/AuE (Au Electrode)

Covalent immobilization of a mixture of lipase, GK, and GPO onto nanocomposite of Pin5COOH/AuPPy/AuE is carried out as follow: Add 20 µl mixture of enzymes on the surface of this Pin5COOH/AuPPy modified AuE and keep it at 4 °C. After 10 min, wash the electrode with DW and use it for electrochemical measurements (Narang et al., 2012).

4.2.2.6 Chemistry of Immobilization of Enzymes on Pin5COOH/AuPPy Nanocomposite

The chemical reaction involved in the co-immobilization of enzymes onto Pin5COOH/AuPPy nanocomposite includes the covalent cross-linking between $-NH_2$ groups on surface of enzyme and $-COOH$ group of Pin5COOH/AuPPy nanocomposite through amide linkage as follow:

$$\underset{\text{Enzyme}}{E-NH_2} + HOOC\text{-}5\text{-}Pin/AuPPy \text{ nanocomposite}$$
$$\rightarrow \underset{\text{Immobilized enzyme}}{E-NH-CO-5-Pin/AuPPy} + H_2O$$

The steps involved in immobilization of enzymes on nanocomposite of AuPPy/Pin5COOH are shown in schematic diagram in Fig. 5.

4.2.2.7 Confirmation of Co-Immobilization of Enzymes

The co-immobilization of enzymes onto nanocomposites of ZnONPs-CHIT/Pt and AuPPy/Pin5COOH/Au is confirmed by high-resolution SEM, electrochemical impedance spectroscopic spectra (EIS), FTIR spectra, and cyclic voltammetry (CV) (Autolab, Eco Chemie, The Netherlands. Model: AUT83785) as follow.

SEM: Figure 6A–D shows SEM images of bare Pt electrode, CHIT/Pt, ZnONPs/CHIT/Pt, and enzymes/CHIT/ZnONPs/Pt. The SEM image of bare Pt electrode displayed the homo granular space (Fig. 6A), while CHIT deposited Pt electrode has rough surface due to deposition of CHIT film (Fig. 6B). The ZnONPs/CHIT/Pt electrode showed the dispersed ZnONPs in chitosan matrix (Fig. 6C). Immobilization of enzymes onto

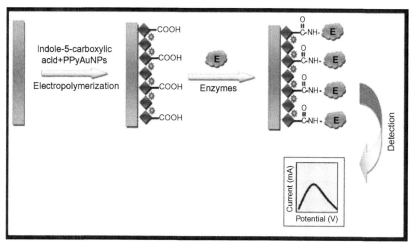

Figure 5 Schematic representation of AuPPy/Pin5COOH electrode preparation and chemical reactions involved in covalent co-immobilization of enzymes (◉, AuPPy; ◆, Pin5COOH). *Reprinted from Narang et al. (2012), with permission from Elsevier.* (See the color plate.)

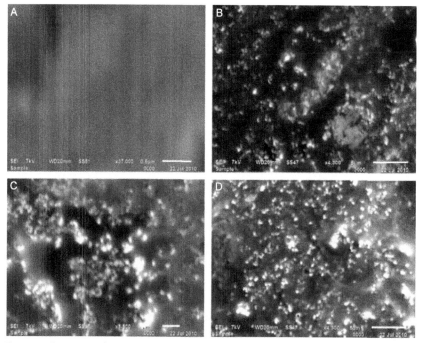

Figure 6 Scanning electron microscopic (SEM) images of (A) bare electrode, (B) CHIT/electrode, (C) ZnONPs/CHIT/electrode, and (D) enzymes/ZnONPs/CHIT/electrode. *Reprinted from Narang and Pundir (2011), with permission from Elsevier.*

the granular morphology of ZnONPs/CHIT nanocomposite changed into the regular form to confirm the interaction between the nanocomposite and enzymes (Fig. 6D) (Narang & Pundir, 2011).

Figure 7A–C exhibits the SEM images of bare Au electrode, Pin5COOH/AuPPy/Au electrode, and enzymes/Pin5COOH/AuPPy/Au

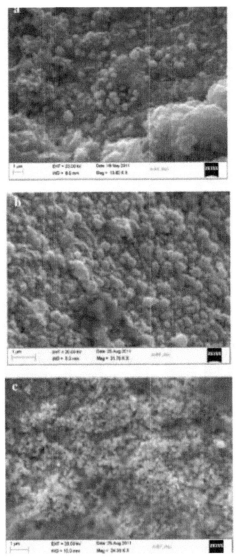

Figure 7 Scanning electron microscopic (SEM) images of (A) bare electrode, (B) Pin5COOH/AuPPy/Au, and (C) enzyme/Pin5COOH/AuPPy/Au. *Reprinted from Narang et al. (2012), with permission from Elsevier.*

electrode The surface morphologies of bare (Au) electrode show smooth surface (Fig. 7A), nanocomposite modified Au electrode exhibit granular morphology with heterogeneous roughness of Pin5COOH/AuPPy/Au electrode depicting the dispersed NPs in Pin5COOH matrix (Fig. 7B), while the enzymes/Pin5COOH/AuPPy/Au electrode, has the granular morphology of Pin5COOH/AuPPy changed into the regular form, due to the co-immobilization/interaction between Pin5COOH-AuPPy and enzymes (Fig. 7C) (Narang et al., 2012).

EIS spectra: Electrochemical impedance spectra also known as dielectric spectroscopy measures the dielectric properties of a medium as a function of frequency. Figure 8 shows spectra of (■) CHIT/Pt electrode, (•) nanoZnO-CHIT/Pt electrode, and (▲) enzymes/nanoZnO-CHIT/Pt electrode. The charge transfer process in enzymes/nanoZnO-CHIT/Pt electrode is studied by monitoring charge transfer resistance (R_{ct}) at the electrode and electrolyte interface. The R_{ct} value depends on the dielectric and insulating features at the electrode/electrolyte interface. The R_{ct} values for the CHIT/Pt, nanoZnO-CHIT/Pt, and enzymes/nanoZnO-CHIT/Pt electrodes as calculated from semicircle of EIS spectra are 6.68×10^2, 4.47×10^2, and 22.38×10^2 Ω, respectively. The electron transfer via redox couple is hindered by the presence of enzymes on electrode surface. The increased R_{ct} value of enzymes/nanoZnO-CHIT/Pt electrode is due to the fact that most biological molecules, including enzymes, are poor electrical conductors at low frequencies (at least <10 KHz) and cause hindrance to the electron transfer (Narang & Pundir, 2011).

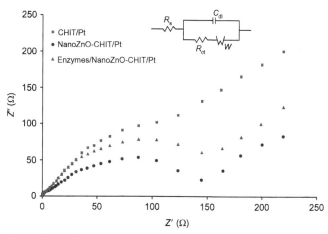

Figure 8 Electrochemical impedance spectra (EIS) of (■) CHIT/Pt, (•) ZnONPs/CHIT/Pt, and (▲) enzymes/ZnONPs/CHIT/electrode. *Reprinted from Narang and Pundir (2011), with permission from Elsevier.*

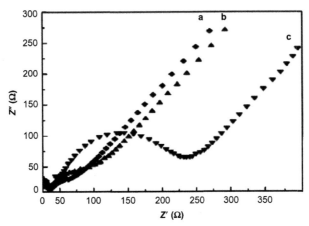

Figure 9 Electrochemical impedance spectra (EIS) of (a) AuPPy/Au, (b) Pin5COOH/AuPPy/Au, and (c) enzyme/Pin5COOH/AuPPy/Au. Reprinted from Narang et al. (2012), with permission from Elsevier.

Figure 9a–c shows EIS of (a) AuPPy/Au electrode, (b) Pin5COOH/AuPPy/Au electrode, and (c) enzymes/Pin5-COOH/AuPPy/Au electrode. The R_{ct} values for the AuPPy/Au electrode, Pin5-COOH/AuPPy/Au electrode, and enzymes/Pin5COOH/AuPPy/Au electrode are 90, 60, and 225 Ω, respectively. The increased R_{ct} value of enzymes/Pin5-COOH/AuPPy/Au electrode is due to the immobilization of enzymes onto Pin5COOH/AuPPy/Au surface (Narang et al., 2012).

FTIR spectra: The FTIR spectra of (i) CHIT/Pt, (ii) ZnONPs-CHIT/Pt electrode, and (iii) enzymes/ZnONPs-CHIT/Pt electrode are compared in Fig. 10. The spectrum of CHIT shows characteristic absorption band of aminosaccharide at 3421, 2811, and 1647 cm^{-1}, due to overlapping of –OH and –NH$_2$ stretchings, CH$_2$ stretching, and C–O stretching, respectively. CHIT-ZnONPs composite films show an additional peak at 492 cm^{-1} corresponding to ZnONPs, which confirms the formation of ZnONPs-CHIT hybrid. Enz/CHIT-ZnONPs/electrode shows broadening of the peak at 3264 and 1653 cm^{-1}, due to the addition of carbonyl groups of glutaraldehyde and amino groups on the surfaces of enzymes, bound onto ZnONPs-CHIT film (Narang & Pundir, 2011).

Figure 11a and b shows the FTIR spectra for Pin5COOH/AuPPy/Au electrode (curve a) and enzymes/Pin5COOH/AuPPy/Au electrode (curve b). The spectrum associated with the oxidized Pin are characterized by a very large adsorption band located in the spectral domain between 3700 and 3100 cm^{-1}, which is characteristic of –OH groups belonging to

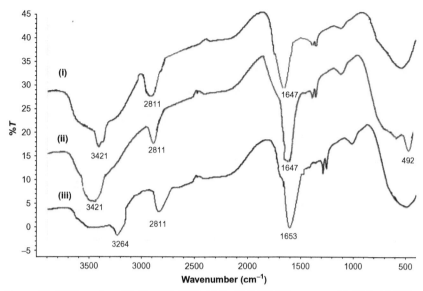

Figure 10 FTIR spectra of (i) CHIT/Pt, (ii) ZnONPs/CHIT/Pt, (iii) enzymes/ZnONPs/CHIT/Pt. *Reprinted from Narang and Pundir (2011), with permission from Elsevier.*

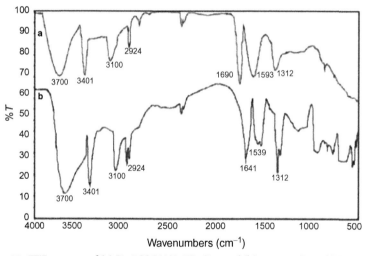

Figure 11 FTIR spectra of (a) Pin5COOH/AuPPy/Au and (b) enzyme/Pin5COOH/AuPPy/Au. *Reprinted from Narang et al. (2012), with permission from Elsevier.*

residual water molecules trapped in the polymer matrix as well as water molecules absorbed in Pin. The peaks at 3401, 2924, 1593, and 1312 cm^{-1} are due to –N–H stretching, –C–H (aromatic) stretching, C–C stretching, and C–N stretching (between two indole units),

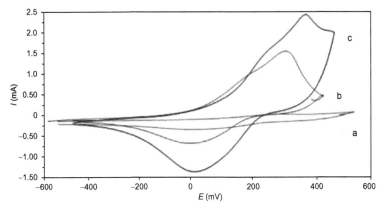

Figure 12 Cyclic voltammograms (a) CHIT/Pt, (b) ZnONPs/CHIT/Pt, (c) enzymes/ZnONPs/CHIT/Pt. Reprinted from Narang and Pundir (2011), with permission from Elsevier.

respectively. A sharp absorption peak at 1690 cm^{-1} is due to the attachment of –COOH group onto the indole. FTIR spectrum of enzymes/Pin5COOH/AuPPy/Au (curve b), confirm the covalent immobilization of enzyme as indicated by the appearance of additional absorption bands at 1641 and 1539 cm^{-1} assigned to the carbonyl stretch (amide I band) and –N–H bending (amide II band), respectively (Narang et al., 2012). Further, enzyme attachment to the nanocomposites is confirmed when the enzymes are not released after washing of the enzyme electrodes with detergents or guanidine hydrochloride solutions.

CV: Figure 12a–c shows the cyclic voltammograms for CHIT/Pt electrode, ZnO-CHIT/Pt electrode, and enzymes/nanoZnO-CHIT/Pt electrode in PB (50 mM, pH 7.0, plus 0.9% NaCl) containing [Fe(CN)$_6$]$^{3]4-}$ (5 mM) recorded at different scan rates (10–100 mVs^{-1}), respectively. The anodic potential was shifted toward positive side and the cathodic peak potential shifted in the reverse direction. The redox potential of nanoZnO-CHIT electrode shifted toward the higher side than that of the pure CHIT film, which is due to the incorporation of ZnONPs in the film. The nanoZnO-CHIT nanocomposite electrode appears to provide a biocompatible environment to the enzymes, and ZnONPs act as electron mediators, resulting in an accelerated electron transfer between enzymes and electrode (Narang & Pundir, 2011).

Cyclic voltammograms for AuPPy/Au electrode, Pin5COOH/AuPPy/Au electrode, and Enz/Pin5COOH/AuPPy/Au electrode in presence of 20 mM triolein in PBS (0.1 M, pH 7.5) are shown in Fig. 13. No peak is observed in case of the bare electrode. The polymer showed two

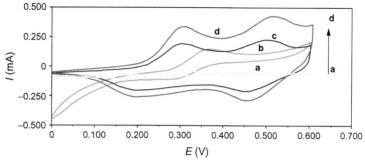

Figure 13 Cyclic voltammograms (CV) of (a) bare electrode, (b) AuPPy/Au, (c) Pin5COOH/AuPPy/Au, and (d) enzyme/Pin5COOH/AuPPy/Au (Narang et al., 2012). *Reprinted from Narang et al. (2012), with permission from Elsevier.*

characteristic peaks, which are the known oxidation and reduction peaks observed in acetonitrile. The nanocomposite provided a conductive path for electron transfer and promoted electron transfer at lower potentials (Narang et al., 2012).

4.3 Kinetic Properties of Co-Immobilized Lipase, GK, and GPO
4.3.1 Optimum pH

Most enzymes have highest activity around an optimum pH (or pH range) and therefore, at other pH values, enzymatic activity decreases substantially. Certain pH values might promote dissociation of multimeric enzymes and decrease their activities. Amino acid side chains in the active site may act as weak acids or/and bases and hence, the catalytic activity depends on the pH of the medium. The pH range over which enzyme activity changes substantially can provide clues as to the type of amino acid residues involved in the catalytic cycle. Binding of the enzyme on a nanosurface might alter the local pH of the medium surrounding the bound enzyme, and hence, this can influence the pH optimum for the bound enzyme to be different from that of the soluble enzyme. The ionized side chains of proteins may also play an important role in the interactions that maintain enzyme/protein structure. To study optimum pH for the activity of the bound enzyme, 0.1 M sodium phosphate buffer was used, which was adjusted to have the pH in the range of 6.0–8.0 at an interval of 0.5. The optimal pH of the mixture of the three soluble enzymes are examined here (pH 7.2) which is increased to 7.5 when bound to ZnONPs/CHIT/Au electrode (Fig. 14) (Narang & Pundir, 2011) and the pH optimum decreased to 6.5 upon binding to AuPPy/Pin5-COOH/Pt electrode (Narang et al., 2012) (Table 1).

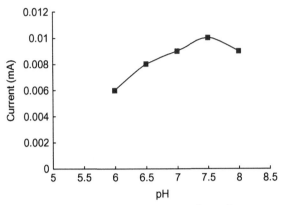

Figure 14 Effect of pH on enzymes/ZnONPs/CHIT/Pt electrode. *Reprinted from Narang and Pundir (2011), with permission from Elsevier.*

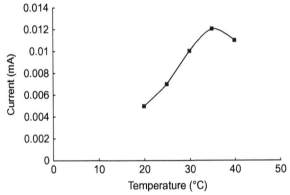

Figure 15 Effect of temperature on enzymes/ZnONPs/CHIT/Pt modified electrode. *Reprinted from Narang and Pundir (2011), with permission from Elsevier.*

4.3.2 Optimum Temperature

In addition to the pH optimum, most enzymes may also have an optimum temperature for maximal activity. The optimum temperature of the both samples of the bound enzymes is 35 °C and this is a small increase when compared to that of the corresponding free enzymes. For example, an increase of 2 °C for lipase when co-immobilized with GK and GPO onto the nanosupport (Fig. 15) (Narang & Pundir, 2011; Narang et al., 2012), but could be well within the experimental error. Co-immobilized enzymes show enhanced thermostability when compared to those of the corresponding free enzymes (data not shown). This increased thermal stability could be due to intramolecular covalent cross-linkages with the support (nanocomposite), which prevent conformational changes in the enzyme at higher temperature, and thus inhibits the enzyme deactivation (Table 1).

4.3.3 Response Time
Another important characteristic for most biocatalysts is their response time, which is defined as the time needed for an enzyme to produce an electrochemical response due to the product generated during the catalytic cycle. Thus, not only the catalytic cycle is involved but the generation of the electrochemical signal from the product at appropriate potential is reflected in the response time. When lipase, GK, and GPO are bound to CHIT/ZnO nanofilm, the current is produced within 6 s (Fig. 16) (Narang & Pundir, 2011). However, when these enzymes are immobilized on AuPPy/Pin5COOH, the response time was 4 s (Narang et al., 2012) which is slightly faster than that of the mixture of the free enzymes (15 min). The later result could be due to faster diffusion of the product to the electrode or that the free enzyme has to diffuse to the electrode surface for the chemistry to occur. Thus, electrode-bound enzymes could be substantially more advantageous for biocatalysis applications (Table 1).

4.3.4 Working Range/Linearity
The concentration range over which the substrate is consumed and product produced linearly is called the working range and it is an important parameter for biocatalysis, as this allows for the most efficient use of the biocatalyst. Above or below this range, there may be excess or insufficient amount of the substrate to take full advantage of the biocatalyst capacity. When lipase, GK,

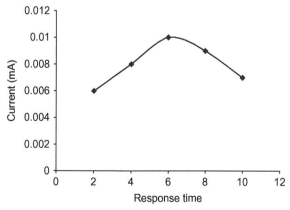

Figure 16 Response time for enzymes/ZnONPs/CHIT/Pt modified electrode. *Reprinted from Narang and Pundir (2011), with permission from Elsevier.*

and GPO were immobilized on CHIT/ZnO nanofilm, the working range has been found to be between 50 and 650 mg/dl (Narang & Pundir, 2011). When the enzymes were immobilized on AuPPy/Pin5COOH nanocomposite, this range was 50–700 mg/dl (Narang et al., 2012). These are comparable to that of the mixture of free enzymes (40–400 mg/dl) (Table 1).

4.3.5 Storage Stability

The storage stability is the time during which enzyme can be stored without considerable loss in its activity, and this is another important criterion for the practical application of biocatalyst. Longer the storage time, the greater the utility of the biocatalytic preparation. The mixture of lipase, GK, and GPO were stored at 4 °C for 6 months, while co-immobilized enzymes on CHIT/ZnONPs (Narang & Pundir, 2011) and AuPPy/Pin5COOH nanocomposite (Narang et al., 2012) have been stored at 4 °C for 7 months without considerable loss in their activities, which is only slightly longer than that of the mixture of free enzymes (6 months) (Table 1).

Table 1 A Comparison of Kinetic Properties of Lipase, Glycerol Kinase, and Glycerol-3-Phosphate Oxidase Co-Immobilized onto Two Different Nanocomposites Electrodeposited on Metal Electrode

Property	Mixture of Free Enzyme	Enz/CHIT/ZnONPs/Pt (Narang & Pundir, 2011)	Enz/Pin5COOH/AuPPyNPs/Au (Narang et al., 2012)
Type of electrode	–	Pt	Au
Method of immobilization	–	Covalent	Covalent
Type of bonding	–	Schiff base	Amide linkage
Optimum pH	7.2	7.5	6.5
Optimum temperature (°C)	37	35	35
Potential for maximum current (V)	–	0.4	0.1
Detection limit for triolein (mg/dl)	–	20	20
Linear range for triolein (mg/dl)	40–400	50–650	50–700
Response time	15 min	6 s	4 s
Storage stability (months)	6	7	7

CHIT, Chitosan; ZnONPs, zinc oxide nanoparticles; Pin5COOH, poly (indole-5-carboxylic acid); AuPPy, gold polypyrrole nanocomposite.

4.4 Application of Co-Immobilized Lipase, GK, and GPO onto Nanomaterials

The co-immobilized lipase, GK, and GPO onto nanomaterials were tested for the construction of improved amperometric biosensor for the detection of TGs in biological fluids. A biosensor is defined as an analytical device which consists of a biological sensing element connected to a transducer which converts biological response into a measurable signal. For practical applications, the magnitude of the signal should be directly proportional to the concentration of analyte over sufficient concentration range. A biosensor has the advantage over other analytical devices in terms of their simplicity, rapidity, sensitivity, and specificity. Among the various types of biosensors known, amperometric biosensors are quite common, which measure current, against the concentration of the analyte.

4.4.1 Principle of Amperometric TG Biosensor

In amperometric TG biosensor, H_2O_2 is generated from TG by the combined reaction of lipase, GK, and GPO in a reaction cascade, as described earlier. The product, H_2O_2, decomposes at high voltages to release electrons as described by the following equation:

$$H_2O_2 \rightarrow 2H^+ + 1/2 O_2 + 2e^-$$

The flow of electrons, i.e., current, is directly proportional to the concentration of hydrogen peroxide produced by the enzyme reactions, which is in turn proportional to the concentration of the analyte (TG).

4.4.2 Fabrication and Response Measurement of an Amperometric TG Biosensor

To construct an amperometric TG biosensor, modified metal electrodes are loaded with enzymes co-immobilized onto nanocomposites, and serve as the working electrode. This electrode along with Ag/AgCl as standard electrode and Pt wire as auxiliary electrode are connected through potentiostat/galvanostat. The three electrode system is immersed into 3 ml of 0.1 M phosphate buffer, pH 7.5. Triolein is added into the reaction buffer and current generated due to the enzymatic reaction is measured under an applied potential. The potential applied to electrode with Enz/ZnONPs-CHIT/PtE CHIT-ZnO nanocomposite was 0.4 V (Narang & Pundir, 2011) and 0.1 V, in the case of Enz/AuPPy/pin5COOH/AuE (Narang et al., 2012).

Table 2 Triglycerides Level of Healthy and Diseased Persons as Measured by TG Biosensor Based on Enzyme/AuPPy/Pin5COOH/Au Modified Electrode

		Serum Triglyceride (mg/dl) (Mean ± S.D.)	
Age Group (n = 8)	Sex	Healthy Persons	Diseased Persons
11–20	M	90.0 ± 10.4	230.0 ± 15.2
	F	88.3 ± 11.6	210.0 ± 10.1
21–30	M	129.2 ± 43.6	394.1 ± 39.5
	F	105.2 ± 22.3	390.0 ± 57.1
31–40	M	136.2 ± 51.2	422.0 ± 68.1
	F	110.3 ± 30.2	420.0 ± 57.2
41–50	M	164.2 ± 54.6	572.8 ± 51.3
	F	142.1 ± 38.9	413.0 ± 35.4
51 and above	M	195.0 ± 43.5	591.6 ± 54.5
	F	184.2 ± 26.1	495.6 ± 58.3
Range	M	90.0–195.0	230–591.6
	F	88.3–184.2	210–495.6

Reprinted from Narang et al. (2012), with permission from Elsevier.

4.4.3 Analytic Use of TG Biosensor

The TG biosensor is applied for amperometric determination of TG in the serum of healthy people and compared with those suffering from hypertriglyceridemia, and the values have been found to be in the range of 88–195 mg/dl and 210–590 mg/dl, respectively (Table 2) (Narang et al., 2012).

4.4.4 Evaluation of TG Biosensor

Enz/ZnONPs-CHIT electrode has a detection limit (LOD) as low as 20 mg/dl, with the working range 50–650 mg/dl, while Enz/AuPPy/pin5COOH/AuE has the same LOD but slightly higher working range of 50–700 mg/dl. The % recovery of added triolein in the serum is 94–96% as measured by Enz/ZnONPs-CHIT/PtE, but 91–95% by Enz/AuPPy/pin5COOH/AuE (Table 3) (Narang & Pundir, 2011). The coefficient of variation (CV) within and between batches are <3% by Enz/ZnONPs-CHIT/PtE (Table 4) (Narang & Pundir, 2011), while the same are 4.14% and 5.85%, respectively, in case of

Table 3 Analytical Recovery of Added Triglyceride In Sera Determined by TG Biosensor Based on ZnONPs

Triglyceride Added (mg/dl)	Triglyceride Found (mg/dl)	% Recovery
–	180	–
20	199.2	96.0 ± 0.23
50	227.3	94.6 ± 0.40

Reprinted from Narang and Pundir (2011), with permission from Elsevier.

Table 4 Within and Between Batch Assay Coefficient of Variation (CV) for the Determination of Serum Triglyceride by TG Biosensor Based on ZnONPs/CHIT/PtE

n	Triglycerides (mg/dl)	CV (%)
Within assay (5)		
110.89	109.61 ± 1.15	1.04
109		
107.99		
110.4		
109.8		
Between assay (5)		
114.4		2.75
115.5	117.1 ± 3.23	
120.9		
114.4		
120.3		

Reprinted from Narang and Pundir (2011), with permission from Elsevier.

Enz/AuPPy/pin5COOH/AuE. A good correlation was observed ($r = 0.98–0.99$) between the serum triglycerides values obtained by the standard enzymic colorimetric method and these biosensors.

The performance of the TG biosensor is unaffected by a number of serum substances at their corresponding physiological concentrations (Narang & Pundir, 2011) as shown in Table 5.

Table 5 Interference Study of Various Serum Compounds on TG Biosensor Based on ZnONPs

Interferents	Relative Response (%)
Glucose	100
Fructose	100
Ethanol	100
Ascorbic acid	100
Citric acid	100
Lactic acid	100
Malic acid	97
Tartaric acid	100
Alanine	100
Leucine	100
Urea	98
Uric acid	95
Cholesterol	90
Bilirubin	90
Creatinine	100

Reprinted from Narang and Pundir (2011), with permission from Elsevier.

REFERENCES

Barbosa, O., Ortiz, C., Berenguer-Murcia, Á., Torres, R., Rodrigues, R. C., & Fernandez-Lafuente, R. (2015). Strategies for the one-step immobilization-purification of enzymes as industrial biocatalysts. *Biotechnology Advances, 33*, 435–456.

Barlett, P. N., Dawnson, D. H., & Farrington, J. (1992). Electrochemically polymerized films of 5- carboxyindole. *Journal of Chemical Society Faraday Transaction, 88*, 2685–2695.

Duff, M., & Kumar, C. V. (2009). Molecular signatures of protein binding to alpha-ZrP: Isothermal titration calorimetric studies. *Journal of Physical Chemistry, 113*, 15083–15089.

Fossati, P., & Prencipe, L. (1982). Serum triglyceride determined colorimetrically with an enzyme that produces hydrogen peroxide. *Clinical Chemistry, 28*, 2077–2080.

Jiang, H. L., Kwon, J. T., Kim, Y. K., Kim, E. M., Arote, R., Jeong, H. J., et al. (2007). Galactosylated poly(ethylene glycol)-chitosan-graft-polyethylenimine as a gene carrier for hepatocyte-targeting. *Gene Therapy, 14*, 1389–1398.

Kalia, V., & Pundir, C. S. (2002). Co-immobilization of lipase, glycerol kinase, glycerol-3-phosphate oxidase and peroxidase onto alkylamine glass beads through glutaraldehyde coupling. *Indian Journal of Biochemistry & Biophysics, 39*, 342–346.

Kazenwadel, F., Franzreb, M., & Rapp, B. E. (2015). Synthetic enzyme supercomplexes: Co-immobilization of enzyme cascades. *Analytical Methods, 7*, 4030–4037.

Kalia, V., & Pundir, C. S. (2004). Evaluation of serum triglyceride determination with lipase, glycerol kinase and glycerol-3- phosphate oxidase and peroxidase co-immobilized onto alkylamine glass beads. *Indian Journal of Biochemistry & Biophysics, 41,* 326–328.

Kumar, C. V., & Chaudhari, A. (2002). High temperature activity of heme proteins at zirconium phosphate. *Chemical Communications, 20,* 2382–2383.

Kumar, C. V., & Chaudhari, A. (2000). Protein immobilized at the galleries of alpha-zirconium phosphates: Structure and activity studies. *Journal of the American Chemical Society, 122,* 830–837.

Lata, S., Batra, B., Singala, N., & Pundir, C. S. (2013). Construction of amperometric L-amino acids biosensor based on L-amino acid oxidase immobilized onto ZnONPs/c-MWCNT/PANI/AuE. *Sensors and Actuators B: Chemical, 188,* 1080–1088.

Liao, J. D., Lin, S. P., & Wu, Y. T. (2005). Dual properties of the deacetylated sites in chitosan for molecular immobilization and biofunctional effects. *Biomacromolecules, 6,* 392–399.

Miao, Y., Wu, X., Chen, J., Liu, J., & Qiu, J. (2008). Preparation of Au/polypyrrole composite nanoparticles and study of their electro-catalytical reduction to oxygen with (without) lacasse. *Gold Bulletin, 41,* 336–340.

Minakshi, & Pundir, C. S. (2008a). Co-immobilization of lipase, glycerol kinase, glycerol-3-phosphate oxidase and peroxidase on to aryl amine glass beads affixed on plastic strip for determination of triglycerides in serum. *Indian Journal of Biochemistry & Biophysics, 42,* 111–115.

Minakshi, & Pundir, C. S. (2008b). Construction of an amperometric enzymic sensor for triglyceride determination. *Sensors Actuators B: Chemical, 133,* 251–255.

Mudhivarthi, V. K., Bhambhani, A., & Kumar, C. V. (2007). Novel enzyme/DNA/inorganic nanomaterials: A new generation of biocatalysts. *Dalton Transactions, 21,* 5483–5497.

Narang, J., & Pundir, C. S. (2011). Construction of triglyceride biosensor based on chitosan-ZnO nanocomposite film. *International Journal of Biological Macromolecules, 49,* 701–705.

Narang, J., Bhambi, M., Minakshi, & Pundir, C. S. (2010). Determination of serum triglyceride by enzyme electrode using covalently immobilized enzyme on egg shell membrane. *International Journal of Biological Macromolecules, 47,* 691–695.

Narang, J., Chauhan, N., Rani, P., & Pundir, C. S. (2012). Construction of an amperometric biosensor based on AuPPy nanocomposite and poly (indole-5-carboxylic acid) modified Au electrode. *Bioprocess and Biosystems Engineering, 36,* 425–432.

Nelson, D. M., Cox, M. M., & Lehninger. (2000). *Principles of biochemistry* (4th ed.). New York: Worth publishing. Chapter 6.

Chowdhury, R., Stromer, B. S., Pokharel, B., & Kumar, C. V. (2012). Control of enzyme-solid interactions by chemical modification. *Langmuir, 28,* 11881–11889.

Novak, M. J., Pattammattel, A., Koshmerl, B., Puglia, M., Williams, C., & Kumar, C. V. (2016). 'Stable-on-the-Table' enzymes: Engineering enzyme-graphene oxide interface for unprecedented stability of the biocatalysts. *ACS Catalysis, 6,* 339–347.

Pundir, C. S., Singh, S., Bharvi, & Narang, J. (2010). Construction of an amperometric triglyceride biosensor using PVA membrane bound enzymes. *Clinical Biochemistry, 43,* 467–472.

Singh, S. P., Arya, S. K., Pandey, M. K., Malhotra, B. D., Saha, S., Sreenivas, K., et al. (2007). Cholesterol biosensor based on rf sputtered zinc oxide nanoporous thin film. *Applied Physics Letters, 91,* 233106.

Vu, Q. T., Pavlik, M., Hebestreit, N., Pfleger, J., Rammelt, U., & Plieth, W. (2005). Electrophoresis deposition of nanocomposites formed from polythiophene abd metal oxides. *Electrochimica Acta, 51,* 1117–1124.

Wang, J. X., Sun, X. W., Wei, A., Lei, Y., Cai, X. P., Li, C. M., et al. (2006). Zinc oxide nanocomb biosensor for glucose detection. *Applied Physics Letters, 88,* 233106.

Wei, A., Sun, X. W., Wang, J. X., Lei, Y., Cai, X. P., Li, C. M., et al. (2006). Enzymatic glucose biosensor based on ZnO nanorod array grown by hydrothermal decomposition. *Applied Physics Letters, 89,* 123902.

Winartasaputra, H., Kutan, S. S., & Cuilbault, G. C. (1982). Amperometric determination of triglyceride in serum. *Analytica Chimica Acta, 54,* 1987–1990.

CHAPTER ELEVEN

BioGraphene: Direct Exfoliation of Graphite in a Kitchen Blender for Enzymology Applications

C.V. Kumar[*,†,‡,§,1], A. Pattammattel[*]

[*]Department of Chemistry, University of Connecticut, Storrs, Connecticut, USA
[†]Department of Molecular and Cell Biology, University of Connecticut, Storrs, Connecticut, USA
[‡]Department of Inorganic and Physical Chemistry, Indian Institute of Science, Bengaluru, Karnataka, India
[§]Institute of Material Science, University of Connecticut, Storrs, Connecticut, USA
[1]Corresponding author: e-mail address: challa.kumar@uconn.edu

Contents

1. Introduction — 226
2. Mechanism of Exfoliation — 230
3. Tunability of the BioGraphene Characteristics — 231
4. Protein Binding to Graphene and Some Biological Applications — 232
5. Methods — 233
   - 5.1 Preparation of bG with BSA — 233
   - 5.2 Characterization of bG/BSA — 234
   - 5.3 Raman Spectroscopy Measurements — 235
   - 5.4 Raman Data Analysis — 235
   - 5.5 Electron Microscopy Studies — 236
   - 5.6 Storage and Stability Studies — 237
   - 5.7 Zeta-Potential Measurements — 237
   - 5.8 UV–Vis Measurements — 238
   - 5.9 Protein Binding to bG — 239
   - 5.10 Activity of HRP Bound to bG — 240
6. Conclusions — 241

Acknowledgments — 242
References — 242

Abstract

A high yielding method for the aqueous exfoliation of graphite crystals to produce high quality graphene nanosheets in a kitchen blender is described here. Bovine serum albumin (BSA), β-lactoglobulin, ovalbumin, lysozyme, and hemoglobin as well as calf serum were used for the exfoliation of graphene. Among these, BSA gave the maximum exfoliation efficiency, exceeding 4 mg mL^{-1} h^{-1} of graphene. Quality of graphene produced was examined by Raman spectroscopy, which indicated 3–5 layer graphene of very high quality and very low levels of defects. Transmission electron microscopy indicated an average size of ~0.5 μm flakes. The graphene/BSA dispersions were stable over pH 3.0–11, and at 5°C or 50°C, for more than 2 months. Current approach gave higher rates of BSA/graphene (BioGraphene) in better yields than other methods. Calf serum, when used in place of BSA, also gave high yields of good quality BioGraphene and these preparations may be of direct use for cell culture studies. A simple example of BioGraphene preparation is described that can be adapted in most laboratories, and graphene-adsorbed glucose oxidase is nearly as active as the free enzyme. Current approach may facilitate large-scale production of graphene in most laboratories around the world and it may open new opportunities for biological applications of graphene.

1. INTRODUCTION

Graphene is a one-atom thick, two-dimensional carbon allotrope, and it is being touted as a wonder material with high promise for a large number of practical applications as well as theoretical studies (Geim & Novoselov, 2007). Being 1000 times stronger than steel, slightly better conducting than silver, light weight, high surface area per unit mass, and with the ability to biodegrade (Bianco, Kostarelos, & Prato, 2011), its promise for biomedical applications is high (Chung et al., 2013; Girish, Sasidharan, Gowd, Nair, & Koyakutty, 2013). To realize this promise of graphene for biological applications, one major problem to overcome is the large-scale production of graphene in aqueous media, under conditions that are suitable for biological systems. For example, light weight and large surface area of graphene will enable high loadings of proteins/enzymes on the nanosheets, as a novel platform for biomedical applications (Chung et al., 2013), enzyme fuel cells (Liu, Alwarappan, Chen, Kong, & Li, 2010), biocatalysis (Novak et al., 2016; Zore, Pattammattel, Gnanaguru, Kumar, & Kasi, 2015), biosensors (Pumera, 2011), and supercapacitors (El-Kady, Strong, Dubin, & Kaner, 2012; Mosa et al., 2016).

Graphene applications in enzymology can combine the advantages of layered materials with the amazing properties of graphene to produce unique substances. Immobilization of enzymes on graphite, which usually requires surface activation, resulted in some denaturation of the enzyme due to unfavorable interactions with the hydrophobic surfaces of graphite (Ianniello, Lindsay, & Yacynych, 1982). Enzymes showed poor retention of activity on graphite surface and the enzyme loadings ranged from 0.65 to 1.3 mg cm^{-2} which is not considered high (Zhou & Chen, 2001). On the other hand, intercalation of enzymes between the nanosheets of 2D materials such as metal phosphates indicated high retention of activities (~90%), protected them from degradation by heat, microbes, or proteases, and indicated significant stability enhancements in specific cases (Kumar & Chaudhari, 2000). Therefore, intercalation of enzyme between the graphene sheets could be interesting and has the potential to improve enzyme stabilities. For example, layered inorganic phosphates were excellent for the binding of a variety of enzymes and resulted in improved physical, chemical, and biochemical characteristics (Deshapriya & Kumar, 2013). One hypothesis is that when the support matrix of a redox enzyme is electrically conducting, the redox reactions catalyzed by the enzyme may be accelerated where the support matrix might facilitate long distance electron transfer between the oxidation and reduction equivalents, even when the partners are far apart. In a similar fashion, high thermal conductivity of the support matrix could facilitate rapid heat dissipation in an exothermic enzymatic reaction and protect the enzyme from thermal denaturation during the catalytic cycle. Such interesting hypotheses can be tested when the support matrix has high electrical and thermal conductivities, as in the case of graphene. Graphene offers an excellent platform for enzyme binding because of its high surface area (700–1000 $m^2\ g^{-1}$), mechanical strength, high electrical/thermal conductivities, as outlined above (Lu, Yang, Zhu, Chen, & Chen, 2009).

Graphite is exfoliated in organic solvents (Yi & Shen, 2015), ionic liquids (Morishita, Okamoto, Katagiri, Matsushita, & Fukumori, 2015) or using surfactant solutions (Lotya, King, Khan, De, & Coleman, 2010) either by sonication or by applying shear. But biocompatibility of these preparations for cell culture, drug delivery, or other biological applications is seriously limited due to the toxicity or incompatibility of the additives or organic solvents used (Yang, Li, Tan, Peng, & Liu, 2013). Therefore, a biocompatible solvent medium with solutes that are also biologically acceptable while using water as the solvent is highly desirable. The method of production should also be simple, accessible in low resource environments, and highly reproducible so that a novice user can readily be able to make the preparations.

Making progress toward a more biocompatible preparation of graphene nanosheets, bovine serum albumin (BSA) was used under ultrasonication to make graphene (Ahadian et al., 2015). BSA was suggested to stabilize and facilitate graphite delamination upon ultrasonication but the underlying mechanism of exfoliation is still under debate (Ahadian et al., 2015). The strongly hydrophobic graphene nanosheets need to be stabilized in the aqueous phase to prevent their restacking and proteins such as BSA could help improve the solubility of the flakes by adsorption on the exposed hydrophobic surfaces of graphene (Laaksonen et al., 2010). However, the ultrasonication method introduces oxidative defects (10–25%) in the exfoliated graphene due to the prolonged exposure to oxygen and high energy density of cavitation induced by the ultrasonic waves (Skaltsas, Ke, Bittencourt, & Tagmatarchis, 2013). Scalability of ultrasonication process to produce gram or kilogram quantities of graphene dispersions is also challenging and only limited exfoliation rates of <1 mg mL^{-1} h^{-1} have been reported by sonication methods (Paton et al., 2014). Breaking of the graphene sheets under sonication has been observed, and these can produce reactive sites at newly created edges while reaction of these active sites with oxygen can produce defect sites or produce undesirable hotspots. Without a facile method of making biocompatible preparations of graphene in water as the preferred medium, the widespread use of graphene for biological applications is limited.

Recently, the shear force/turbulence produced in an ordinary kitchen blender has been used for the mechanical exfoliation of graphite in the presence of organic solvents (Yi & Shen, 2014) or surfactants (Varrla et al., 2014). The shear-exfoliation, as opposed to ultrasonication, generates a turbulent flow which exceeds the critical Reynolds number required for delamination (10^4) of the graphene flakes (Varrla et al., 2014), and even a simple kitchen blender can generate enough shear force for this purpose. Without proper approach for the stabilization of the released nanosheets, however, graphene sheets will restack and there will not be net exfoliation or the efficiency of exfoliation could be low. Thus, a stabilizing agent is required to improve the efficiency and stabilize the flakes, but the stabilizing agent needs to be compatible with the intended biological applications, or else the corresponding preparations will not be useful for these sensitive applications (Table 1).

Solvent-exfoliated or surfactant-exfoliated preparations, for example, are not suitable for immediate biological applications which may interfere with the components in the cell culture media or toxic to the sensitive biological systems. Therefore, a simple, biologically compatible method for producing large quantities of graphene/water dispersions is urgently required. We combined the excellent characteristics of BSA to stabilize hydrophobic surfaces with the simplicity of using an ordinary kitchen blender to produce protein-coated graphene for applications in enzymology.

Previous studies on graphene oxide showed that BSA coating enhanced its affinity for proteins and stabilized it further (Pattammattel et al., 2013). BSA is an inexpensive common protein, available in large quantities as a waste product from the meat industry. BSA is also a good choice because of its ability to bind to a variety of hydrophobic ligands, drugs, lipids,

Table 1 Comparison of Graphene Exfoliation Efficiencies (mg mL^{-1} h^{-1}) with the Current Method

Method	Efficiency	pH and Temperature
Ultrasonication/Gelatin/water (Ge, Wang, Shi, & Yin, 2012)	0.038	7.0 and 50°C
Ultrasonication/BSA/water (Ahadian et al., 2015)	0.27	3.6 and 50°C
Ultrasonication/hydrophobins/60% ethanol (Gravagnuolo et al., 2015)	0.75	NA and 25°C
Ultrasonication/polysaccharides/water (Unalan, Wan, Trabattoni, Piergiovanni, & Farris, 2015)	1	NA and 25°C
Shear/detergent (Varrla et al., 2014)	0.2	NA and 25°C
Shear/DMF (Yi & Shen, 2015)	0.4	NA and 25°C
Shear/NMP (Paton et al., 2014)	0.0025	NA and 25°C
Shear/BSA (current method) (Pattammattel & Kumar, 2015)	4	7.0 and 25°C

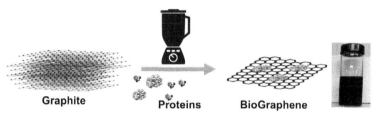

Scheme 1 Production of BioGraphene using proteins in a kitchen blender. (See the color plate.)

hormones, peptides, and a variety of substrates (Kumar & Buranaprapuk, 1997). BSA coating on the graphene surface could also inhibit the formation of protein corona, which is often found when nanoparticles are introduced into the circulation of mammals. Therefore, serum albumin turns out to be a good choice for these studies to test the above hypothesis of graphite exfoliation with proteins when subjected to shear forces.

The protein-blender method described here gave high concentrations of nearly defect-free graphene aqueous dispersions in less than an hour of blending (Scheme 1). Along these lines, several ordinary proteins were also found to facilitate the exfoliation of graphite to graphene at comparable rates, and these included readily available proteins such as β-lactoglobulin, ovalbumin, lysozyme, glucose oxidase, and even hemoglobin. These proteins are also expected to be biocompatible and sourced from food or fermentation industry (edible proteins) (Yun-HwaPeggy & JackAppiah, 2011). This protein-based method is also attractive over the use of organic solvents or surfactants, and a variety of different proteins can be chosen for the graphene production, further enhancing the tailoring of the BioGraphene (bG) preparations for specific biological applications.

For comparison purposes, the shear-based exfoliation is more favorable over ultrasonication, as the former can be readily scaled-up in most laboratories without expensive instrumentation. Several liters of graphene/water dispersions of 4–7 mg mL^{-1} (w/v) were readily produced in our lab, in less than a day. Such scalability coupled with flexibility of using a variety of biomolecules for exfoliation is essential for a wider application of graphene, particularly in the biological world, where the requirements of the experiments could vastly vary from system to system.

2. MECHANISM OF EXFOLIATION

Brief discussion of a rudimentary mechanism of exfoliation is essential before proceeding, while recognizing that the mechanism of this process could be much more complex than illustrated here. However, any insight gained even with a rudimentary mechanism could be useful to tailor the method for special applications by individual users. Due to the inherent hydrophobic nature of the graphene surface, which is made of sheets of sp^2 carbon atoms, the sheets are to be solubilized in water by reducing their surface energy. Two primary free energy components need to be considered. One is the energy needed to separate the plates from the stacks against van der Waals interactions between the plates, which is positive free energy. The other major energy component will be the solvation energy provided by the interaction of the dissolved protein with the graphene nanosheet displacing the water molecules in the surroundings. The sum of these two free energy components should be large and negative for the exfoliation process to proceed. One can note that the delamination energy is characteristic of the 2D material and independent of the protein used while the solvation energy provided by the protein could strongly depend on the structure and nature of the protein employed. Thus, careful selection of the protein is recommended in designing the shear-based exfoliation experiments, as variations on the theme described here.

3. TUNABILITY OF THE BioGraphene CHARACTERISTICS

Based on the above tentative mechanism of exfoliation, one could test specific hypotheses and produce particular variations of the bioGraphene (bG) production. BSA turned out to be an excellent protein to assist in the exfoliation and stabilization of graphene suspensions but a number of other proteins also work well (Pattammattel & Kumar, 2015). A strong correlation between the exfoliation efficiency and the magnitude of negative charge on the protein has been demonstrated from our data but there are still subtle variations among the negatively charged proteins that have been examined by us. Graphene concentrations as high as 7 mg mL^{-1} are readily produced by our method, and the exfoliation rate is related to a number of experimental parameters such as nature of the protein used, blade speed, protein concentration, pH of the medium, and temperature, which can be controlled systematically.

One advantage of using different proteins for the exfoliation is that the isoelectric point of the resulting bG can be readily tuned. The isoelectric point of the bG appears to be identical to that of the protein used, and hence, using proteins, the preparations can be stabilized at particular pH and solution conditions, as desired by the end user. Coating the graphene nanosheets with desired proteins is also valuable in the rational control of how the bG interacts with the target biological systems. For example, graphene coated with a particular protein could be of interest for a given application than BSA. This feature of tailoring the nature of the protein in the current method will be of particular value to the individual users who can tailor their preparations based on specific attributes of their experimental needs.

In summary, we tested the ability of some simple proteins to aid in graphite exfoliation. Thus, energy is needed to peel off the graphene sheets, which is supplied by the appropriate shear forces applied to the surface of the graphite crystals (Paton et al., 2014). Energy is also supplied by the favorable interaction of selected proteins with graphene and the overall free energy change governs the propensity for the process. The shear-based method, in conjunction with proteins as stabilizing agents, avoids the use of organic media or surfactants which can denature delicate enzymes to be used or adversely influence sensitive biological systems such as cells or cell membranes and other components where the graphene needs to be utilized.

4. PROTEIN BINDING TO GRAPHENE AND SOME BIOLOGICAL APPLICATIONS

There are a few of examples of studies of protein or enzyme binding to graphene, and the rarity of examples is mainly because of the restricted availability of this wonder material for the biological community. Consequently, the fundamental understanding of how biomolecular systems interact with graphene from thermodynamic and kinetic considerations is a serious gap in our current knowledge. For example, hydrophobins were used to exfoliate graphite, but quantitation of protein affinities, enzyme stabilities, or thermodynamic driving force for the binding or mechanism of binding to graphene is still not well understood (Gravagnuolo et al., 2015). In our studies of graphene exfoliation, the rate of exfoliation depended on protein charge, irrespective of protein size, or protein hydrophobicity index. Protein–graphene interactions are complex, not understood, but they are central in controlling protein-mediated graphene exfoliation or for the application of bG and this understanding is also important for the rational development of graphitic materials for biological applications. Along these lines, some examples of the use of graphene in the biological world are presented below, but these are only few of the very many reported in the literature.

Graphene films made by chemical vapor deposition (CVD) were found to accelerate the growth of human mesenchymal stem cells and their differentiation in a controlled manner, suggesting high potential of graphene for regeneration therapy (Nayak et al., 2011). In another study, graphene grown on silicon, glass, polystyrene, or polydimethoxysilane by CVD were compared as supports for human stem cell growth. Controlled cell growth and cell proliferation improved consistently with graphene coatings when compared to the respective bare surfaces of the corresponding underlying substrates (Kalbacova, Broz, Kong, & Kalbac, 2010). Preconcentration of specific cell growth inducers at the graphene surface by noncovalent binding was suggested to be one factor in promoting cell growth (Lee et al., 2011). Graphene coatings also supported neurite growth in mouse hippocampal cell culture model (Lin et al., 2011). Neurite numbers and average length increased significantly after cell deposition on graphene when compared to growth on the corresponding tissue culture polystyrene. These studies clearly establish the potential of graphene for biomedical applications.

Furthermore, the flat, hydrophobic surface of graphene is readily amenable for surface modification. For example, laminin-coated graphene films enhanced differentiation of human neural stem cells to neurons when compared to glass substrates (Park et al., 2011). Few layer graphene (FLG) generated by electric arc was dispersed in chitosan solutions and addition of 0.1–0.3 wt.% of graphene increased the elastic modulus of graphene–chitosan films by 200% (Fan et al., 2010). L929 cells adhered and developed on graphene–chitosan composites and demonstrated its good biocompatibility.

Graphene is also being examined as a nanoplatform for sensing. In one study, pyrene-3-butyric acid was adsorbed onto graphene and its COOH group has been covalently linked to the amino groups on the lysine side chains of pepsin to attach the enzyme onto graphene nanosheets (Centeno & Elorza, 2015). The large surface area of graphene allowed for high enzyme loading on graphene, also attachment in specific patterns on graphene surface. Graphene is also being tested for drug delivery applications with success, but more studies are urgently needed to further establish its utility for therapeutic purposes (Liu, Robinson, Tabakman, Yang, & Dai, 2011).

5. METHODS

5.1 Preparation of bG with BSA

Equipment:
1. Regular kitchen blender (Oster) with 1.2 L cup or other size.
2. Desktop centrifuge (Fisher).
3. UV–Vis spectrophotometer (HP/Waters).

Materials:
1. Graphite flakes (+100 mesh, Sigma–Aldrich, MO).
2. Bovine serum albumin (Equitech).
3. Sodium phosphate buffer (Sigma).

Preparation method:
1. BSA was dissolved in 200 mL deionized (DI) water and pH was adjusted to 7.0 using dilute HCl or NaOH.
2. Graphite (20 g) was added to the above into the blender flask.
3. Graphite is exfoliated using the shear force in a kitchen blender. The above mixture of graphite and BSA was subjected to blending at ~17,000 rpm speed for 90 min. Blending for 30 s and rest for 30 s cycle is recommended to avoid excessive heating of the mixture.

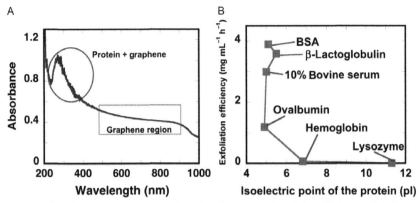

Figure 1 (A) Absorption spectrum of BioGraphene. (B) Plot of the efficiency of different proteins in graphene production versus isoelectric point of the protein.

4. Unexfoliated graphite is separated from the mixture by centrifugation. The mixture was transferred to 400 mL plastic bottles and centrifuged at 1500 rpm for 45 min. The supernatant carefully separated from unexfoliated graphite pellet.
5. The concentration of the graphene in the dispersion can be estimated using UV–vis spectrophotometer. Record the extinction at 660 nm, with appropriated dilution using DI water (Figure 1A). Graphene concentration of ~6 mg mL^{-1} are readily achieved by this protocol (extinction coefficient: 4 mg^{-1} mL cm^{-1}) (Pattammattel and Kumar, 2015).

Tip 1: If the mixture is excessively heating, cool it by keeping inside the fridge (~4°C).

Tip 2: If you wish to monitor the progress of exfoliation, aliquots of 500 μL samples can be taken out at specific intervals.

Tip 3: Graphene may be produced using other negatively charged proteins, at neutral pH, instead of BSA. A few are listed here.

The efficiency of different proteins in exfoliating graphite is showed in Figure 1B. The charge of the protein was found to be a critical factor in determining the exfoliation efficiency. Surprisingly, lysozyme, a positively charged protein (+11 at pH 7.0) showed no observable exfoliation whereas all the negatively charged proteins were effective in the exfoliation process.

5.2 Characterization of bG/BSA

Raman spectroscopy and transmission electron microscopy (TEM) are used to characterize graphene produced (Ferrari & Basko, 2013). Raman spectroscopy can quantify the number of layers, size, and defect type of the graphene (Paton et al., 2014). TEM images can confirm the size of graphene produced and a qualitative analysis of number of layers of graphene present in a given sample.

5.3 Raman Spectroscopy Measurements

Equipment:
1. Laser Raman Spectrometer (Reinshaw 514 nm laser source).

Materials:
1. Microscope cover glass (Fisher Scientific).

Procedure:
1. Graphene suspension (50 µL, 6 mg mL^{-1}) was dried on a regular glass surface to get ~0.5 cm diameter spot size.
2. Raman spectra of the deposited graphene is collected using a 514 nm laser source from 3200 to 1200 cm^{-1} range. Control experiment was performed using parent graphite flakes.
3. Characteristic peaks were positioned around 2700 cm^{-1} (2D), 1580 cm^{-1} (G), and 1350 cm^{-1} (D).

5.4 Raman Data Analysis

The peak position, intensity, and shape of the peaks depend on the size, number of layers, and defects present in the graphene sample. The peak characteristics were compared with that of parent graphite. After exfoliation, major changes occur at the 2D region and the D region. The asymmetrical 2D band of graphite was centered around 2725 cm^{-1}, and as a result of exfoliation, the peak became symmetrical and shifted to 2695 cm^{-1} (Figure 2A).

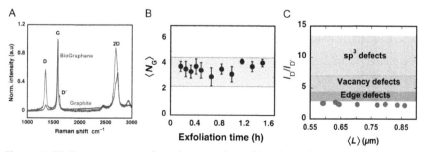

Figure 2 (A) Raman spectra of graphene and graphite show changes in the intensity and position of characteristic bands. (B) Raman analysis of the data indicated the average flake size and average number of layers in the flake. (C) Samples show minor edge defects, as analyzed by Raman spectra.

This change in the positions of the peak and shoulder is used to quantify the number of layers of graphene present in the sample, using the empirical Eq. (1) (Paton et al., 2014).

$$\langle N_G \rangle = 10^{0.84M + 0.45M^2}; \quad M = \frac{(I_{2D}/I_{2Ds})_{Graphene}}{(I_{2D}/I_{2Ds})_{Graphite}} \quad (1)$$

N_G, number of layers; M, constant; I_{2D} – Intensity corresponding to the 2D band (2725 cm^{-1} and I_{2Ds} – Intensity corresponding to the shoulder of the 2D band (~2695 cm^{-1}).

> *Tip*: About 100 separate spectra are needed to make statistically useful quantitative analysis by Raman spectroscopy. An excel sheet can be conveniently used to use above equation for analysis of such large data sets.

The changes in 2D region after exfoliation represents the defects and edges of the sheets, and they are used to quantitate the lateral size of graphene sheets and the type of defects present in graphene samples (Eq. 2):

$$\langle L \rangle = \frac{k}{\left\{ [I_D/I_G]_{Graphene} - [I_D/I_G]_{Graphite} \right\}} \quad (2)$$

L is lateral size of the sheets; I_D and I_G are Intensities of D band (1350 cm^{-1}) and G band (1583 cm^{-1}), respectively; and k, 0.17 (experimentally calculated; Paton et al., 2014).

> *Tip*: The described synthetic protocol used to result ~0.5 μm graphene 2–4 layers, using the above analysis of the Raman data (Figure 2A and B).

5.5 Electron Microscopy Studies

A qualitative analysis of exfoliation can be carried out from TEM images of graphene.

Equipment:
1. FEI Tecnai T12 TEM

Materials:
1. Cu-grids (Ted Pella Inc., CA)

Procedure:
1. Graphene dispersion in water (1 μL, 20 μg mL^{-1}) was drop casted to a Cu-grid and dried under vacuum for 3–4 h.

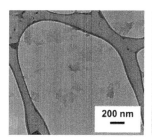  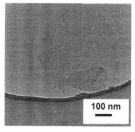

Figure 3 Representative TEM images of micron size, few layer, and BSA-coated BioGraphene.

2. The Cu-grid can be directly taken for imaging at accelerating voltage of 100 kV.
3. After acquiring the images, they are analyzed using ImageJ software (from NIH, free software) to estimate the average size of the sheets.
4. TEM analysis of our samples confirmed the presence of <1 μm size few layer graphene (Figure 3).

  Tip: Selected area electron diffraction pattern of the sheets can be done to confirm the number of layers or crystallinity of graphene.

5.6 Storage and Stability Studies

The colloidal stability of graphene dispersion is an important parameter to be analyzed for biological applications. Poor stability of the dispersion will result in settling down or precipitation, induced by aggregation of the particles. The stability of the solutions is carried out using zeta-potential studies and UV–Vis spectroscopy.

5.7 Zeta-Potential Measurements

Equipment:
1. Brookhaven zeta plus zeta-potential analyzer (Holtsville, NY).

Materials:
1. Polystyrene cuvette (2 mL, 1 cm path length) (Sigma).
2. Dilute NaOH and dilute HCl (Sigma).

Procedure:
1. Graphene suspension (1.6 mL, ~1 mg mL^{-1}) is taken in a polystyrene cuvette and pH was adjusted to 7.0 by the addition of dilute acid or base, as needed.
2. Electrode for the zeta-potential analyzer immersed in the suspension and connected to the instrument.

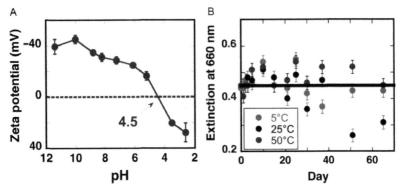

Figure 4 (A) Change in zeta potential of BioGraphene with pH. (B) Absorbance measurements show stability over 2 months at 5°C or 50°C. (See the color plate.)

3. The instrument uses laser Doppler velocimetry to calculate the zeta potential using Smoluchowski fit.
4. The graphene suspensions should give ~-25 mV charge at pH 7.0 (Figure 4A).
5. The zeta potential of the dispersion can be analyzed as a function of pH using dil. NaOH or dil. HCl.

5.8 UV–Vis Measurements

Extinction of graphene platelets at 660 nm can be used to monitor the dispersion stability of graphene at different pHs, temperatures, and also the storage stability at different temperatures.

Equipment:
1. UV–Vis spectrophotometer (Agilent Technologies, Santa Clara, CA).

Materials:
1. Glass cuvette (1 cm path length) (Sigma).
2. Dil. NaOH and dil. HCl (Sigma).

Procedure:
1. Dilute graphene dispersion in water is taken in a 500-μL cuvette and 660 nm absorbance is noted.
2. For temperature stability studies, the temperature of the dispersion was varied from 25°C to 80°C, using the in-built thermostat in the instrument and absorbance at 660 nm was noted.
3. For pH stability studies, the pH of the dispersion was varied from 2 to 10 using NaOH or HCl and absorbance at 660 nm was recorded.

4. For storage stability studies, ~20 mL graphene dispersion (6 mg mL^{-1}, pH 7.0) kept at three different temperatures (4°C, 25°C, and 50°C). The absorbance values of the solutions were recorded everyday for 60 days (Figure 4B).

Both zeta-potential studies and absorbance measurements clearly proved the stability of graphene dispersions at biologically relevant conditions for over a month (Figure 4B). Zeta-potential studies indicated the isoelectric point around 4.5 and the dispersions indicated low stability around pH 5.0. Storage stability of the dispersions depended on the temperature. At room temperature (25°C), half-life was 1 month, while at 5°C and 50°C, the dispersions were stable for much longer, and there has been no change in sample absorbance over 60 days at these latter temperatures (Figure 4B).

5.9 Protein Binding to bG

One of the applications of bG could be for use in biocatalysis or biosensing. Hence, we tested the binding of a model enzyme to graphene, and examined its enzymatic activity. A number of proteins are tested in our lab for binding to the bG preparations and several of them showed high loadings, while some indicated moderate loadings. For example, RNAse A and Catalase showed high loadings (80–90%, w/w), while hemoglobin and horseradish peroxidase (HRP) indicated only moderate loadings (35–40%, w/w). The solution pH, buffer concentration, or other solution conditions are not optimized and hence, the binding could be improved further, in a systematic manner. Also the binding of the enzyme could be different depending on the preparation of the graphene via different routes. The size, number of layers, and defects can be different depending on the preparation conditions such as type of mechanical force used (sonication, shear, etc.), pH, temperature, and type of exfoliating agent (proteins, surfactants, carbohydrates, etc.) Here, HRP bound to the bG retained excellent activity which further supports the notion that this material could be of high value for enzymology applications. We describe below the binding of HRP as a model enzyme due to its widespread use in biosensing and in enzymology (Figure 5).

Figure 5 HRP binding to the nanosheets of BioGraphene for sensing or biocatalysis applications. (See the color plate.)

Equipment:
1. UV–Vis spectrometer (Agilent Technologies, Santa Clara, CA).

Materials:
1. Guaiacol (Sigma).
2. Hydrogen peroxide (Fisher Scientific).
3. Horseradish peroxidase (Sigma).

Procedure:
1. Prepare stock solutions of HRP (~5 mg mL^{-1}) in 10 mM Na$_2$HPO$_4$ buffer at pH 7.0 and measure the molar concentrations using absorption spectroscopy (Soret absorbance at 403 nm).
2. Prepare a series of HRP/bG mixtures with increasing concentrations of the enzyme in 2 mL eppendorf tubes. Typically, graphene concentration was fixed at 3 mg mL^{-1} but that of HRP increased, 0.5, 1.0, and 2.0 mg mL^{-1}.
3. To 500 µL of 10 mM Na$_2$HPO$_4$ buffer at pH 7.0 buffer containing desired concentration of HRP, add 500 µL of bG (6 mg mL^{-1}). The mixture equilibrated for 8 h.
4. The sample was centrifuged at 8000 rpm for 90 min to separate HRP/graphene complex and the unbound HRP is quantified using its Soret absorbance at 403 nm.
5. Loading of HRP to graphene, under the above conditions, was about 40–50% (w/w) in our experiments.

   Tip 1: Shaking the sample during incubation will help to equilibrate the mixture faster.

   Tip 2: Since graphene shows absorption at 280 nm, estimation of protein concentration at that wavelength might lead to large errors. However, Soret absorbance of HRP at 403 nm is reliable, and hence, we suggest using the latter.

5.10 Activity of HRP Bound to bG

The peroxidase activity of the enzyme bound to bG was determined calorimetrically using H$_2$O$_2$ as a substrate and guaiacol as a chromogenic reactant. The data show that the initial rates of the biocatalyst are essentially the same as that of the free enzyme, which clearly shows that the bound enzyme is fully active. Activities of HRP bound to graphene in comparison to unbound enzyme are described below (Figure 6).

Equipment:
1. UV–Vis spectrometer (Agilent Technologies, Santa Clara, CA).

Materials:
1. Guaiacol (Aldrich).
2. Hydrogen peroxide (1 mM, Aldrich).

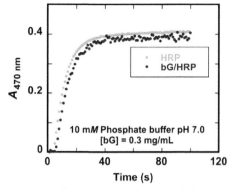

Figure 6 Peroxidase activity of HRP bound to bG (red dots, lower set in the print version) using guaiacol as the substrate when compared to that of the free enzyme (blue dots, upper set in the print version). The initial rates are nearly the same for both samples.

3. Horseradish peroxidase (Sigma).
4. Phosphate buffer (10 mM phosphate buffer at pH 7.0) (Sigma).

Procedure:
1. Prepare the bG/HRP sample as described above with initial concentration of HRP as 5 μM (∼1 mg mL^{-1}), and centrifuge the sample to remove unbound HRP (∼2 μM).
2. Resuspend the pellet in 1 mL 10 mM phosphate buffer at pH 7.0.
3. Prepare stock solutions of H_2O_2 (25 mM) and guaiacol (75 mM) in phosphate buffer.
4. Mix bG/HRP (0.25 μM enzyme), 100 μM guaiacol, and buffer (to make up to a total volume of 2 mL) in a 4-mL cuvette.
5. Set the spectrophotometer to monitor absorbance at 470 nm for 120 s, with a steady stirring inside the cuvette and add 0.5 mM H_2O_2.
6. The initial increase in absorbance was used to calculate the specific activity of the enzyme (typically 3–20 s).
7. Specific activity of free HRP was 0.16 μM s^{-1} and that of bG/HRP was 0.14 μM s^{-1}.

Tip: Sonication of guaiacol in water might be required to prepare its stock solution.

6. CONCLUSIONS

The protein-stabilized graphene suspensions are stable against temperature, pH, and storage time, under biologically relevant conditions and they could be good alternatives to the widely used GO or rGO for biological applications (Ferrari et al., 2015). The bG generated here is amenable to chemical or biochemical modifications and such changes can be made either on the BSA coating or on the graphene surface itself. Thus, the bG reported here is expected to be versatile for a wide use of these nanosheets. This simple exfoliation method could be readily replicated in most laboratories, even in resource-limited regions, this rational approach might lead to a revolution in the use of graphene for not only biological applications but also other applications of this wonder material.

In this report, we present a simple and convenient method to produce high quality graphene dispersions that are suitable for biological applications. The graphene dispersions are stable for long times, even when stored at room temperature, under normal conditions. Storage at 4°C enhanced the storage time and this could be because of the inhibition of bacterial growth in the samples. The storage may be improved by the addition of antibacterial agents but we wanted to demonstrate the high stability of bG, as prepared in the laboratory. The preparations are amenable for enzymology applications which are yet to be fully explored.

ACKNOWLEDGMENTS

We thank the Fulbright foundation for a fellowship given to C.V.K. to visit the IISc, India, and financial support from the University of Connecticut Research Excellence Award, and the EAGER Award from the National Science Foundation (DMR-1441879).

REFERENCES

Ahadian, S., Estili, M., Surya, V. J., Ramon-Azcon, J., Liang, X., Shiku, H., et al. (2015). Facile and green production of aqueous graphene dispersions for biomedical applications. *Nanoscale, 7*, 6436–6443.

Bianco, A., Kostarelos, K., & Prato, M. (2011). Making carbon nanotubes biocompatible and biodegradable. *Chemical Communications, 47*, 10182–10188.

Centeno, A., & Elorza, A. Z. (2015). Graphene in biotechnology. *Material Matters, 8*, 25.

Chung, C., Kim, Y. K., Shin, D., Ryoo, S. R., Hong, B. H., & Min, D. H. (2013). Biomedical applications of graphene and graphene oxide. *Accounts of Chemical Research, 46*, 2211–2224.

Deshapriya, I. K., & Kumar, C. V. (2013). Nanobio interfaces: Charge control of enzyme/inorganic interfaces for advanced biocatalysis. *Langmuir, 29*, 14001–14016.

El-Kady, M. F., Strong, V., Dubin, S., & Kaner, R. B. (2012). Laser scribing of high-performance and flexible graphene-based electrochemical capacitors. *Science, 335*, 1326–1330.

Fan, H., Wang, L., Zhao, K., Li, N., Shi, Z., Ge, Z., et al. (2010). Fabrication, properties and biocompatibility of graphene-reinforced chitosan composites. *Biomacromolecules, 11*, 2345–2351.

Ferrari, A. C., & Basko, D. M. (2013). Raman spectroscopy as a versatile tool for studying the properties of graphene. *Nature Nanotechnology, 8*, 235–246.

Ferrari, A. C., Bonaccorso, F., Fal'ko, V., Novoselov, K. S., Roche, S., Boggild, P., et al. (2015). Science and technology roadmap for graphene, related two-dimensional crystals, and hybrid systems. *Nanoscale, 7*, 4598–4810.

Ge, Y., Wang, J., Shi, Z., & Yin, J. (2012). Gelatin-assisted fabrication of water-dispersible graphene and its inorganic analogues. *Journal of Materials Chemistry, 22*, 17619–17624.

Geim, A. K., & Novoselov, K. S. (2007). The rise of graphene. *Nature Materials, 6*, 183–191.

Girish, C. M., Sasidharan, A., Gowd, G. S., Nair, S., & Koyakutty, M. (2013). Confocal Raman imaging study showing macrophage mediated biodegradation of graphene in vivo. *Advanced Healthcare Materials, 2*, 1489–1500.

Gravagnuolo, A. M., Morales-Narváez, E., Longobardi, S., da Silva, E. T., Giardina, P., & Merkoçi, A. (2015). In situ production of biofunctionalized few-layer defect-free microsheets of graphene. *Advanced Functional Materials, 25*, 2771–2779.

Ianniello, R. M., Lindsay, T. J., & Yacynych, A. M. (1982). Differential pulse voltammetric study of direct electron transfer in glucose oxidase chemically modified graphite electrodes. *Analytical Chemistry, 54*, 1098–1101.

Kalbacova, M., Broz, A., Kong, J., & Kalbac, M. (2010). Graphene substrates promote adherence of human osteoblasts and mesenchymal stromal cells. *Carbon, 48*(15), 4323–4329.

Kumar, C. V., & Buranaprapuk, A. (1997). Site-specific photocleavage of proteins. *Angewandte Chemie International Edition, 36*, 2085–2087.

Kumar, C. V., & Chaudhari, A. (2000). Proteins immobilize at the galleries of alpha-zirconium phosphate: Structure and activity studies. *Journal of the American Chemical Society, 122*, 830–837.

Laaksonen, P., Kainlauri, M., Laaksonen, T., Shchepetov, A., Jiang, H., Ahopelto, J., et al. (2010). Interfacial engineering by proteins: Exfoliation and functionalization of graphene by hydrophobins. *Angewandte Chemie International Edition, 49*, 4946–4949.

Lee, W. C., Lim, C. H., Shi, H., Tang, L. A., Wang, Y., Lim, C. T., et al. (2011). Origin of enhanced stem cell growth and differentiation on graphene and graphene oxide. *ACS Nano, 5*, 7334–7431.

Lin, N., Zhang, X., Song, Q., Su, R., Zhang, Q., Kong, T., et al. (2011). The promotion of neurite sprouting and outgrowth of mouse hippocampal cells in culture by graphene substrates. *Biomaterials, 3*, 9374–9382.

Liu, C., Alwarappan, S., Chen, Z., Kong, X., & Li, C.-Z. (2010). Membraneless enzymatic biofuel cells based on graphene nanosheets. *Biosensors and Bioelectronics, 25*, 1829–1833.

Liu, Z., Robinson, J. T., Tabakman, S. M., Yang, K., & Dai, H. (2011). Carbon materials for drug delivery and cancer therapy. *Materials Today, 14*, 316–323.

Lotya, M., King, P. J., Khan, U., De, S., & Coleman, J. N. (2010). High-concentration, surfactant-stabilized graphene dispersions. *ACS Nano, 4*, 3155–3162.

Lu, C.-H., Yang, H.-H., Zhu, C.-L., Chen, X., & Chen, G.-N. (2009). A graphene platform for sensing biomolecules. *Angewandte Chemie International Edition, 121*(26), 4879–4881.

Morishita, T., Okamoto, H., Katagiri, Y., Matsushita, M., & Fukumori, K. (2015). High-yield ionic liquid-promoted synthesis of boron nitride nanosheets by direct exfoliation. *Chemical Communications, 15*, 12068–12071.

Mosa, I., Pattammattel, A., Kadimisetty, K., El-Kady, M. F., Pande, P., Bishop, G. W., et al. (2016). Ultrathin graphene-protein supercapacitors for implantable biomedical devices. *Science Advances*. Under review.

Nayak, T. R., Andersen, H., Makarm, V. S., Khaw, C., Bae, S., Xu, X., et al. (2011). Graphene for controlled and accelerated osteogenic differentiation of human mesenchymal stem cells. *ACS Nano, 5*, 4670–4678.

Novak, M. J., Pattammattel, A., Koshmerl, B., Puglia, M., Williams, C., & Kumar, C. V. (2016). "Stable-on-the-table" enzymes: Engineering the enzyme–graphene oxide interface for unprecedented kinetic stability of the biocatalyst. *ACS Catalysis, 6*, 339–347.

Park, S. Y., Park, J., Sim, S. H., Sung, M. G., Kim, K. S., Hong, B. H., et al. (2011). Enhanced differentiation of human neural stem cells into neurons on graphene. *Advanced Materials, 23*, H253–H257.

Paton, K. R., Varrla, E., Backes, C., Smith, R. J., Khan, U., O'Neill, A., et al. (2014). Scalable production of large quantities of defect-free few-layer graphene by shear exfoliation in liquids. *Nature Materials, 13*, 624–630.

Pattammattel, A., & Kumar, C. V. (2015). Kitchen chemistry 101: Multigram production of high quality biographene in a blender with edible proteins. *Advanced Functional Materials, 25*, 7088–7098.

Pattammattel, A., Puglia, M., Chakraborty, S., Deshapriya, I. K., Dutta, P. K., & Kumar, C. V. (2013). Tuning the activities and structures of enzymes bound to graphene oxide with a protein glue. *Langmuir, 29*, 15643–15654.

Pumera, M. (2011). Graphene in biosensing. *Materials Today, 14*, 308–315.

Skaltsas, T., Ke, X., Bittencourt, C., & Tagmatarchis, N. (2013). Ultrasonication induces oxygenated species and defects onto exfoliated graphene. *The Journal of Physical Chemistry C, 117*, 23272–23278.

Unalan, I. U., Wan, C., Trabattoni, S., Piergiovanni, L., & Farris, S. (2015). Polysaccharide-assisted rapid exfoliation of graphite platelets into high quality water-dispersible graphene sheets. *RSC Advances, 5*, 26482–26490.

Varrla, E., Paton, K. R., Backes, C., Harvey, A., Smith, R. J., McCauley, J., et al. (2014). Turbulence-assisted shear exfoliation of graphene using household detergent and a kitchen blender. *Nanoscale, 6*, 11810–11819.

Yang, K., Li, Y., Tan, X., Peng, R., & Liu, Z. (2013). Behavior and toxicity of graphene and its functionalized derivatives in biological systems. *Small, 9*, 1492–1503.

Yi, M., & Shen, Z. (2014). Kitchen blender for producing high-quality few-layer graphene. *Carbon, 78*, 622–626.

Yi, M., & Shen, Z. (2015). A review on mechanical exfoliation for the scalable production of graphene. *Journal of Materials Chemistry A, 3*, 11700–11715.

Yun-HwaPeggy, H., & JackAppiah, O. (2011). Food-grade proteins from animal by-products. *Handbook of analysis of edible animal by-products* (pp. 13–35). CRCnetBASE.

Zhou, Q. Z. K., & Chen, X. D. (2001). Immobilization of β-galactosidase on graphite surface by glutaraldehyde. *Journal of Food Engineering, 48*, 69–74.

Zore, O. V., Pattammattel, A., Gnanaguru, S., Kumar, C. V., & Kasi, R. M. (2015). Bienzyme–polymer–graphene oxide quaternary hybrid biocatalysts: Efficient substrate channeling under chemically and thermally denaturing conditions. *ACS Catalysis, 5*, 4979–4988.

AUTHOR INDEX

Note: Page numbers followed by "*f*" indicate figures, and "*t*" indicate tables.

A

Abdelmouleh, M., 187
Abe, K., 181–182
Abian, O., 75–76, 79–82
Aboagye, D., 181
Adlercreutz, P., 3–5
Agasti, S.S., 5
Aggarwal, V., 198–221
Aguirre, S.D., 182–183
Ahadian, S., 227–228, 229*t*
Ahopelto, J., 227–228
Ai, H., 180
Ai, Q., 93–94, 107
Algar, W.R., 20–23, 36–37, 39–44, 46–47, 136, 138–139
Ali, M.M., 182–183
Alkasir, R.S.J., 191
Álvaro, G., 181
Alwarappan, S., 226
Anastasova, S., 181–182
Ancona, M.G., 20–23, 26*f*, 31
Andersen, H., 232
Andler, S.M., 2–16, 56–57
Andreescu, D., 183–187, 185*f*, 190
Andreescu, S., 178–191, 185*f*
Anguish, L., 22, 180–181
Ansari, A.A., 183–184
Ansorge-Schumacher, M., 2–3
Anuganti, M., 58, 182–183, 191
Aragón, C.C., 68, 70*f*
Ardao, I., 181
Armisén, P., 2–3, 56–57, 63–64, 66, 75, 79–82
Armstrong, A., 40–41
Arote, R., 201
Arroyo, M., 2–3
Arruda, D.L., 180
Arya, S.K., 201
Asati, A., 185–186
Ayub, M.A.Z., 2–3

B

Babko, A.K., 185–186
Backes, C., 227–228, 229*t*, 231, 234–236
Bae, S., 232
Bale, S.S., 136
Ban, T., 118–119
Bang, A., 3–5
Banks, C.E., 182–183
Barbosa, O., 74–84, 199–200
Barlett, P.N., 207–208
Barnes, C.H.W., 136–139
Baron, R., 180
Barondeau, D.P., 22–23
Barrow, C.J., 152
Basko, D.M., 234–235
Bastida, A., 2–3, 56–57, 63–64, 66
Batalla, P., 83
Batra, B., 203
Bawendi, M.G., 50
Bax, A., 36–37
Becker, R., 143–144
Beelen, T.P., 91
Beganski, C., 152
Belgacem, M.N., 187
Belova, A.B., 114
Benaiges, M.D., 181
Benkovic, S.J., 88
Berenguer-Murcia, Á., 75, 88, 199–200
Bergmeyer, H.U., 143–144
Bernal, C., 75–76
Bernedo, M., 75–76, 79–82
Berti, L., 20–21, 36–37, 136, 138–139
Best, R.B., 36–37
Betancor, L., 58, 75, 152
Bhambhani, A., 22, 56–57, 152, 199–200
Bhambi, M., 200
Bhargava, N., 183–184
Bharvi, 200
Bhirde, A.A., 136–138
Bhushan, A., 136
Bianco, A., 226

Bieth, J., 23–24
Bigley, A.N., 22–23
Bishop, G.W., 136–138, 146–148, 226
Bittencourt, C., 227–228
Blanco-Canosa, J.B., 21, 25–29, 36–37, 39–43, 47
Boeneman, K., 25, 41
Boggild, P., 241–242
Bohlmann, G., 152
Bolivar, J.M., 57–58, 83
Bonaccorso, F., 241–242
Bornscheuer, U.T., 88
Bostancioglu, K., 40
Boufi, S., 187
Bouthillier, F., 74
Bowers, G.N., 30
Bradley, R., 183–184
Brasil, M.C., 2–3
Braun, P.V., 89–90
Braun, R., 144–146
Breger, J.C., 20–50, 26f
Bromley, S.T., 183–184
Brook, M.A., 182–183
Brown, C.W., 21
Broz, A., 232
Brunel, F.M., 32
Bulbul, G., 183–184
Buranaprapuk, A., 228–229
Butte, M.J., 182–183
Byun, S., 2–3

C

Cai, X.P., 201
Caiazzo, R.J., 180
Cargill, A.A., 21, 181
Carrasquillo, K.G., 152
Carrilho, E., 182–183
Carrillo-Carrion, C., 20–21
Caruntu, D., 7
Caruso, F., 89–90
Casey, B.J., 20–21, 36–37, 136, 138–139
Cass, A.E.G., 184
Castillo, J.J., 75–76
Centeno, A., 233
Chailapakul, O., 182–183
Chakraborty, S., 152, 228–229
Chan, S.W., 185–186
Chang, E.L., 32

Chang, S.W., 5
Chang, Y., 42
Chapman, T., 183–184
Chaudhari, A., 22, 56–57, 114–115, 152, 159–162, 169, 199–200, 226–227
Chauhan, N., 202, 206, 207f, 208, 209–210f, 210–220, 212–213f, 215f, 218t, 220t
Chen, G.-N., 226–227
Chen, J., 183–184, 206
Chen, M., 89
Chen, X.D., 114–115, 136–138, 182–183, 226–227
Chen, Z., 226
Cheong, S.C., 138, 142, 146–147
Chernov, M.M., 184
Chi, Q., 114
Chi, X., 47
Chikkaveeraiah, B.V., 136–139, 141–142, 145–147
Chinnapen, D.J.-F., 141–142
Cho, J., 42
Choi, M.J., 42
Choquette, L., 138, 146–148
Chowdhury, R., 159–160, 170, 199–200
Christena, L.R., 180–181
Christensen, R.G., 30
Chung, C., 226
Citterio, D., 181–182
Clapp, A.R., 25, 32, 36–37, 50
Claussen, J.C., 21, 47, 181
Clearfield, A., 158–159, 162
Cohen, R., 22, 180–181
Cole, K.S., 22
Cole-Hamilton, D.J., 152
Coleman, J.N., 227
Collart, F.R., 36–37
Collier, P., 183–184
Comenge, J., 181
Cornish-Bowden, A., 29, 46–47
Costa, A.P.O., 2–3
Costantino, H.R., 152
Cox, M.M., 198–199
Craik, C.S., 23–24
Crivat, G., 179
Cuenca, E., 75–76, 79–82
Cui, Y., 89
Cuilbault, G.C., 200

Currie, M., 42
Curto, V.F., 181–182
Czeslik, C., 180

D

D'Arcy, A., 74
da Silva, E.T., 229t, 232
Dai, H., 233
Das, M., 183–184
Dasgupta, R., 182–183
Datta, S., 180–181
Davies, L.M., 182–183
Davis, V.T., 185–186
Dawnson, D.H., 207–208
Dawson, P.E., 25–29, 36–37, 41–43
De Aberasturi, D.J., 20–21
de Hoog, H.P.M., 89
De, S., 227
Decher, G., 56–57, 157, 159
Deelder, A.M., 58
Delamora-Ortiz, G., 58
Delehanty, J.B., 23, 25, 36–37, 40–42, 49
Dellatore, S.M., 91–92
Demin, S., 180
Deng, N., 157
Deng, Y., 20
Deschamps, J.R., 21, 25, 36–37, 39, 41, 47
Deshapriya, I.K., 20–22, 56–57, 88, 120, 152–174, 226–229
Dhayal, M., 184
Díaz, S.A., 20–50
Dickey, M.D., 182–183
DiCosimo, R., 152
Dimicoli, J.-L., 23–24
Ding, M., 159
Ding, S., 21, 181
Ding, X., 179
Doan, N., 136–138
Doan-Nguyen, V., 183–184
Dordick, J.S., 152
Dorrestein, P.C., 141–142
dos Santos, J.C.S., 74–84
Dowding, J.M., 183–184
Drauz, K., 88
Dubin, S., 226
Duff, M.R., 155, 164–165, 199–200
Duggleby, R.G., 50
Duncan, B., 2–5, 16, 56–57

Dungchai, W., 182–183
Durrant, J.R., 184
Dutta, P.K., 152, 183–184, 228–229
Dwyer, C.L., 20
Dyal, A., 5

E

Eaton, W.A., 36–37
Ebina, Y., 116, 130–131
Eckstein, M., 2–3
Eden, H.S., 136–138
Eggert, T., 74
Elghanian, R., 182–183
El-Kady, M.F., 226
Elorza, A.Z., 233
Ernst, R.D., 183–184
Estili, M., 227–228, 229t
Eyring, E.M., 183–184

F

Fal'ko, V., 241–242
Fan, H., 233
Faria, R.C., 137–138, 146–147
Farrell, D., 38
Farrington, J., 207–208
Farris, S., 229t
Fatemi, F.K., 42
Favre-Bonvin, G., 40
Fernández-Lafuente, R., 2–3, 16, 74–84, 88, 152, 199–200
Fernández-Lorente, G., 16, 56–70, 68–69f, 74–76, 79–82, 152
Ferrari, A.C., 234–235, 241–242
Filice, M., 75
Firouzabadian, L., 152
Fisher, B., 36–37, 50
Fitzpatrick, J., 144–146
Fossati, P., 202–203
Fox, N., 22–23
Fragola, A., 42
Franzreb, M., 199–200
Frasco, T., 183–184, 186–187, 190
Freire, D.M.G., 74–75, 79–82
Fruk, L., 126–127
Fuentes, M., 58, 74–76, 79–82
Fukumori, K., 227

G

Gabig-Ciminska, M., 179
Gandini, A., 187
Ganek, J., 47
Ganesana, M., 184
Gao, C., 89–90
Gao, L., 22, 180–181
Gao, Q., 114–115
Garcia, M.E., 157
Gardossi, L., 88
Gartner, F.H., 144–146
Gassman, N.R., 36–37
Ge, J., 114
Ge, Y., 229t
Ge, Z., 233
Geim, A.K., 226
Gemmill, K.B., 20–21, 23, 36–37, 49, 136, 138–139
Gerrow, K., 141–142
Giardina, P., 229t, 232
Gieskes, W.W., 91
Gijs, M.A., 136–138
Girish, C.M., 226
Gnanaguru, S., 226
Goddard, J.M., 2–16
Godoy, C., 58, 68, 68–69f, 83
Goldman, E.R., 20–21, 25, 36–37
Gong, J.D., 138
Gong, J.L., 182–183
González-Cortés, A., 180
Gopich, I.V., 36–37
Gowd, G.S., 226
Gower, L.B., 90–91
Gowers, S.A., 181–182
Grandy, M.R., 141–142
Grate, J.W., 22, 180
Gravagnuolo, A.M., 229t, 232
Grochulski, P., 74
Groger, H., 88
Gross, R.A., 5, 88
Gu, L., 120–122
Gu, M.B., 2–3
Gu, S., 2–3
Guisán, J.M., 2–3, 16, 56–70, 68–70f, 74–75, 79–82, 152
Guo, H., 42
Gutkind, J.S., 20, 138, 141, 145–147

H

Hall, E.A.H., 180
Hameed, A., 3–5
Hammes-Schiffer, S., 88
Hanefeld, U., 88
Harada, M., 116
Harmer, J., 183–184
Harrison, D., 74
Hart, O., 58, 182–183, 191
Hartono, S.B., 180
Harvey, A., 228, 229t
Harvey, S.P., 22–23
Hasan, F., 3–5
Hastman, D.A., 21
Hauer, B., 152
Hayat, A., 183–184
Hayes, S.A., 185–186
He, Q., 89
Hebestreit, N., 207
Heckert, E., 183–184
Heller, A., 181
Hemmatian, Z., 20
Henry, C.S., 182–183
Hermanson, G.T., 144–146
Hildebrand, M., 91
Hildebrandt, N., 21, 43
Ho, K.M., 120–122
Ho, S.O., 36–37
Ho, Y.P., 23
Hoa, L.Q., 180
Hofele, R.V., 39
Hogeland, K., 152
Homaei, A.A., 180–181
Hong, B.H., 226, 233
Hooks, D.E., 157
Hossain, S.M.Z., 182–183
Houssein, S.M., 182–183
Howarth, M., 141–142
Hu, J., 47
Hu, Y., 179
Huguet, J., 2–3
Hühn, D., 20–21
Huisman, G.W., 88
Hunziker, W., 74
Huston, A.L., 40–41
Hwang, E.T., 2–3
Hyodo, T., 127–128

I

Ianniello, R.M., 226–227
Ikemoto, H., 114
Ikeuchi, T., 180
Illanes, A., 75–76
Illas, F., 183–184
Ionescu, A., 136–139
Ispas, C.R., 179, 184
Israelachvili, J.N., 92
Ito, D., 114–133

J

JackAppiah, O., 229–230
Jaeger, K.E., 74
Jain, V., 25, 40–41
Janak Kamil, Y.V.A., 23–24
Jang, H.S., 181
Jares-Erijman, E.A., 43
Jayathilaka, D., 183–184
Jennings, T.L., 25–29, 36–37, 41–43
Jensen, G.C., 138
Jeon, B.W., 2–3
Jeong, H.J., 201
Jeong, Y., 2–5, 16, 56–57
Jepsen, M.L., 23
Jia, H., 180
Jiang, H., 227–228
Jiang, H.L., 201
Jiang, Y., 88–110
Jiang, Z., 88–110
Jitianu, A., 181
Jo, D.H., 180
Johnson, B.J., 20–22, 25, 40–41
Johnson, S.A., 157
Jones, S., 180
Joo, J.C., 2–3
Josberger, E.E., 20
Joshi, A.A., 137–138
Jovin, T.M., 43
Ju, H., 138–139
Jun, C., 2–3

K

Kadimisetty, K., 137–138, 226
Kainlauri, M., 227–228
Kaittanis, C., 185–186
Kalbac, M., 232
Kalbacova, M., 232
Kalia, V., 200
Kalra, B., 88
Kamada, K., 114–133, 128f
Kamiuchi, M., 114–133
Kampouris, D.K., 182–183
Kaneko, S., 131
Kaner, R.B., 226
Kang, I.K., 42
Kang, J.-F., 183–184
Kang, Y., 183–184
Karakoti, A.S., 183–184
Karimi, A., 178–191
Kaschak, D.M., 157
Kasi, R.M., 58, 182–183, 191, 226
Katagiri, Y., 227
Kautz, R., 20
Kazenwadel, F., 199–200
Kazlauskas, R.J., 88
Ke, X., 227–228
Keene, S., 20
Keesey, J., 143–144
Kelleher, N.L., 141–142
Keller, S.W., 157, 159
Khalid, S., 185–186
Khan, R., 184
Khan, U., 227–228, 229t, 231, 234–236
Khaw, C., 232
Khmelnitsky, Y.L., 114
Kiener, A., 152
Kim, C., 2–3, 5
Kim, C.S., 20, 120, 164
Kim, E.M., 201
Kim, H.-N., 157, 159
Kim, I.S., 42
Kim, J.H., 22, 180
Kim, K.S., 233
Kim, S., 42
Kim, S.N., 138
Kim, Y., 36–37
Kim, Y.H., 2–3
Kim, Y.K., 201, 226
King, P.J., 227
Kipphut, W., 22
Klyachko, N.L., 114
Knudsen, B.R., 23

Koberstein, J., 185–186
Koeleman, C.A.M., 58
Kong, J., 232
Kong, T., 232
Kong, X., 226
Konnert, J.H., 50
Konry, T., 136
Korlann, Y., 36–37
Koshmerl, B., 61, 199–200, 226
Kostarelos, K., 226
Koyakutty, M., 226
Krause, C.E., 136–148
Kreft, O., 89–90, 93–94
Krishnan, S., 137–138
Krohn, R.I., 144–146
Kubo, T., 130–131
Kumachev, A., 182–183
Kumar, A., 88
Kumar, C.V., 20–22, 56–58, 61, 88, 114–115, 120, 137–138, 152–174, 199–200, 226–242, 229t
Küngas, R., 183–184
Kutan, S.S., 200
Kwon, J.T., 201

L

Laaksonen, P., 227–228
Laaksonen, T., 227–228
Lakowicz, J.R., 43, 50
Lalla, R.V., 138, 146–148
Landorf, E.V., 36–37
Lata, J.P., 22, 180–181
Lata, S., 203
Latus, A., 138, 146–147
Le, Q.A.T., 2–3
Lebert, J.M., 182–183
Lee, B.P., 92
Lee, H., 91–92
Lee, N.H., 137–138
Lee, S., 42
Lee, T.G., 180
Lee, W.C., 232
Lehninger, 198–199
Lei, J., 138–139
Lei, Y., 201
Leiter, J.C., 184
Letsinger, R.L., 182–183
Levashov, A.V., 114

Li, C.M., 201
Li, C.-Z., 226
Li, G., 183–184
Li, J., 89, 93–94, 97, 103, 105
Li, N., 233
Li, P., 120–122
Li, W., 183–184
Li, Y., 74, 227
Li, Y.F., 182–184
Liang, X., 227–228, 229t
Liao, J.D., 201
Lim, C.H., 232
Lim, C.T., 232
Lin, N., 232
Lin, S.P., 201
Lin, Y., 58, 182–183, 191
Lindsay, T.J., 226–227
Liu, C., 226
Liu, H., 7
Liu, H.-W., 36–37
Liu, J.W., 180, 182–183, 206
Liu, Y.L., 184
Liu, Z., 114, 227, 233
Liu, Z.M., 184
Llandro, J., 136–139
Long, L., 97
Longobardi, S., 229t, 232
Longworth, S., 183–184
Loos, K., 5
López-Gallego, F., 57–58, 68, 68–69f, 83
Lopez-Vela, D., 75
Lotya, M., 227
Lu, C.-H., 226–227
Lu, H., 159
Lu, Y., 182–183
Luckarift, H.R., 88, 152
Luckham, R.E., 182–183
Lutz, S., 88
Lvov, Y., 180

M

Mabey, D., 179
Madison, E.L., 23–24
Magner, E., 88
Makarm, V.S., 232
Malanoski, A.P., 20–22, 47
Malhotra, B.D., 183–184, 201
Malhotra, R., 138, 142, 146–147

Malik, A., 3–5
Malla, S., 137–138
Mallia, A.K., 144–146
Mallouk, T.E., 157, 159
Malonoski, A.P., 21, 39, 47
Mani, V., 136–139, 141, 145–147
Manoel, E.A., 74–75, 79–82
Manthe, R.L., 36–37
Mao, X., 120–122
Marsh, A., 183–184
Martinez, A.W., 182–183
Martins-DeOliveira, S., 56–70
Mateo, C., 16, 61, 68, 70f, 74–76, 79–82, 152
Matsushita, M., 227
Mattoussi, H., 20–21, 25, 36–38, 50
Mauro, J.M., 36–37, 50
May, O., 88
Mazumdar, D., 182–183
McAuliffe, J., 152
McCauley, J., 228, 229t
McComb, R.B., 30
McGhee, J.D., 165
McLean, L.R., 165–166
McLendon, G.L., 157
Medintz, I.L., 20–50, 181
Mei, B.C., 36–37
Meng, R., 106–107
Merchant, K.A., 36–37
Merk, V., 20–21
Merkoçi, A., 229t, 232
Messersmith, P.B., 91–92
Messmer, B., 89
Metters, J.P., 182–183
Miao, Y., 206
Migani, A., 183–184
Miller, G.L., 67
Miller, W.M., 91–92
Min, D.H., 226
Minakshi, 200
Mink, D., 152, 180
Mirica, K.A., 182–183
Mirkin, C.A., 182–183
Mistri, D., 58, 182–183, 191
Miyake, T., 20
Mohite, D., 3–5
Möhwald, H., 89–90, 93–94
Moore, J.C., 88

Morales-Narváez, E., 229t, 232
Moreno-Perez, S., 56–70
Morgan, N.Y., 136–138
Morishita, T., 227
Moriyasu, A., 127–128, 128f
Morse, D.E., 91
Mosa, I., 226
Mozhev, V.V., 114
Mucic, R.C., 182–183
Mudhivarthi, V.K., 22, 56–57, 152, 199–200
Mudhuvarthi, V.K., 22
Mukai, C., 22, 180–181
Munge, B.S., 138, 142, 146–147
Munilla, R., 56–70
Muñoz, G., 16, 75
Murase, H., 130–131

N

Naffin, J.L., 157
Nagy, A., 23
Naik, R.R., 88
Nair, S., 226
Nakahira, A., 130–131
Nakamura, T., 126–127
Nakayama, A., 118–119
Nallani, M., 89
Narang, J., 200, 202–206, 204f, 206–207f, 208–221, 209–217f, 218t, 220–222t
Naruse, J., 180
Nath, S., 185–186
Naumann, M., 2–3
Nayak, T.R., 232
Nektar Advanced PEGylation, 60
Nelson, D.M., 198–199
Nelson, J.L., 22, 180–181
Neyman, K.M., 183–184
Ng, A.H.C., 148
Nguyen, C., 180
Nicholas, S., 3–5
Nie, Z.H., 182–183
Nieguth, R., 2–3
Niemeyer, C.M., 126–127
Nijhuis, C.A., 182–183
Njagi, J., 179, 184
Nolte, R.J., 89
Noto, M., 5
Novak, M.J., 22, 61, 120, 164, 199–200, 226
Novoselov, K.S., 226, 241–242

O

O'Connor, C.J., 7
O'Neill, A., 227–228, 229t, 231, 234–236
Ofir, Y., 5
Oh, E., 20–23, 25, 26f, 31, 36–37, 39–42, 47, 49, 136, 138–139
Ohya, T., 118–119
Ohya, Y., 118–119
Okamoto, H., 227
O'Keefe, M.J., 185–186
O'Keefe, T.J., 185–186
O'Regan, B., 184
Ornatska, M., 183–187, 185f
Orrego, A.H., 56–70
Ortac, I., 89
Ortiz, C., 74–84, 88, 199–200
Osada, M., 130–131
Othman, A., 178–191
Otieno, B.A., 136–148
Ozel, R.E., 183–184, 191

P

Pacchioni, G., 183–184
Page, M.J., 23–24
Pal, S., 183–184
Palfreyman, J.J., 136–139
Palomo, J.M., 16, 74–75, 152
Pan, F., 89
Pande, P., 226
Pandey, M.K., 201
Papadimitrakopoulos, F., 136–138
Park, J., 233
Park, K., 2–3
Park, M., 2–3, 5
Park, S.Y., 233
Patel, V., 20, 138, 141–142, 145–147
Patil, S., 183–184
Paton, K.R., 227–228, 229t, 231, 234–236
Patra, D., 5
Pattammattel, A., 61, 152, 159–160, 170, 199–200, 226–242, 229t
Pavlik, M., 207
Peczuh, M.W., 137–138
Peeling, R.W., 179
Pelt, S.v., 2–3
Pelton, J.T., 165–166
Pelton, R.H., 182–183

Peng, R., 227
Perez, J.M., 185–186
Perkins, M.D., 179
Peters, R.J., 89
Petrov, A.I., 93–94
Petryayeva, E., 23, 25–29, 36–37, 40–43, 46–47
Pfeiffer, C., 20–21
Pfleger, J., 207
Phadke, G., 138, 146–148
Phillips, S.T., 182–183
Piergiovanni, L., 229t
Pierson, R., 42
Pinaud, F., 49
Pingarrón, J.M., 180
Pirmohamed, T., 183–184
Pizarro, C., 75
Plieth, W., 207
Pokharel, B., 199–200
Polyak, B., 136
Pons, T., 36–37, 42
Popat, A., 180
Poppe, J.K., 2–3
Poulose, A.J., 152
Prasuhn, D.E., 25, 36–37, 41
Prato, M., 226
Preda, G., 183–184
Prencipe, L., 202–203
Prevot, M., 89–90, 93–94
Price, A.D., 89–90
Provenzano, M.D., 144–146
Puglia, M., 61, 152, 199–200, 226, 228–229
Pumera, M., 226
Pundir, C.S., 198–221, 204f, 206–207f, 209–217f, 218t, 220–222t
Puntes, V.F., 181
Puri, M., 152
Pütter, J., 143–144

Q

Qiao, S.Z., 103, 105, 180
Qing Lu, G., 180
Qiu, J., 206
Qu, H., 7

R

Radhakumary, C.S.K., 182–183
Rajaram, Y.R.S., 180–181

Rajendran, V., 126–127
Rameshwar, T., 42
Rammelt, U., 207
Ramon-Azcon, J., 227–228, 229t
Rani, P., 202, 206, 207f, 208, 209–210f, 210–220, 212–213f, 215f, 218t, 220t
Rapp, B.E., 199–200
Rasooly, A., 38
Raushel, F.M., 22–23
Reches, M., 182–183
Reetz, M.T., 74
Rehbock, C., 20–21
Renaud, A., 23–24
Renneberg, R., 89
Riccardi, C.M., 58, 182–183, 191
Riedel, L.M., 183–184
Robins, K., 88
Robinson, J.T., 233
Rocha-Martin, J., 57
Roche, S., 241–242
Rodrigues, D.S., 83
Rodrigues, R.C., 2–3, 75, 83, 88, 199–200
Romero-Fernández, M., 56–70
Rossner, A., 191
Rotello, V.M., 2–16, 56–57
Roth, K.M., 91
Rueda, N., 74–84
Ruhaak, L.R., 58
Ruiz-Matute, A.I., 68, 70f
Rusling, J.F., 136–148
Ryoo, S.R., 226

S

Sabuquillo, P., 2–3, 56–57, 61, 63–64, 66, 75, 79–82
Saha, S., 201
Saito, M., 180
Salon, M.C.B., 187
Samal, S., 42
Samanta, A., 20
Samanta, B., 5
Sánchez-Montero, J.M., 2–3
Santra, S., 185–186
Sapsford, K.E., 20–21, 23, 36–38, 47, 136, 138–139
Sardesai, N.P., 137–138, 183–184, 190
Sariri, R., 180–181
Sasaki, T., 116, 130–131

Sasidharan, A., 226
Schaffer, R., 30
Scherer, N.F., 91–92
Schlenoff, J.B., 56–57
Schmid, A., 152
Schmid, R.D., 74
Schmitt, J., 157, 159
Schoemaker, H.E., 152, 180
Schrag, J.D., 74
Schuler, B., 36–37
Seal, S., 183–184
Seehra, M.S., 183–184
Seker, E., 136
Seneci, C.A., 181–182
Sergeeva, M.V., 114
Shafi, K.V.P.M., 5
Shah, A.A., 3–5
Shahidi, F., 23–24
Sharpe, E., 183–187, 185f, 190
Shchepetov, A., 227–228
Sheldon, R.A., 2–3, 88, 190
Shen, G.L., 184
Shen, K., 136
Shen, M., 89
Shen, Z., 227–228, 229t
Shi, H., 232
Shi, J., 88–110, 114–115
Shi, X., 89
Shi, Y., 183–184
Shi, Z., 229t, 233
Shiku, H., 227–228, 229t
Shimizu, Y., 127–128
Shin, D., 226
Sim, S.H., 233
Simberg, D., 89
Singala, N., 203
Singh, S.P., 183–184, 200–201
Sinisterra, J.V., 2–3
Skaltsas, T., 227–228
Smith, A.M., 182–183
Smith, G.D., 158–159, 162
Smith, P., 74
Smith, P.K., 144–146
Smith, R.J., 227–228, 229t, 231, 234–236
Soh, N., 114–133, 128f
Solanki, P.R., 183–184
Song, L., 159
Song, Q., 232

Song, X., 89–90, 93–94, 107
Spagnoli, C., 5
Spain, J.C., 88
Spengler, M., 126–127
Sreenivas, K., 201
Srinivasan, B., 179
Stahr, F., 180
Stanciu, L.A., 181
Stanish, I., 50
Steenvoorden, E., 58
Stevanato, R., 180–181
Steven Sun, S., 38
Stewart, M.H., 21–23, 25, 26f, 31, 36–37, 39–41, 47
Stoffer, J.O., 185–186
Stone, M.O., 88
Storhoff, J.J., 182–183
Stromer, B.S., 20, 199–200
Strong, V., 226
Su, R., 232
Suescun, A., 75–76
Sugano, Y., 180
Sukhorukov, G.B., 89–90, 93–94
Sun, M., 47
Sun, Q., 93–94, 97
Sun, X., 93–94, 97
Sun, X.W., 201
Sung, M.G., 233
Surya, V.J., 227–228, 229t
Susumu, K., 20–23, 25–29, 26f, 31, 36–37, 39–43, 47, 49
Suzuki, K., 181–182

T

Tabakman, S.M., 233
Tagmatarchis, N., 227–228
Takada, K., 130–131
Takahashi, Y., 118–119
Talbert, J.N., 2–5, 16, 56–57
Talpasanu, I., 180
Tan, X., 227
Tang, L.A., 232
Tekin, H.C., 136–138
Thompson, J.S., 185–186
Thum, O., 2–3
Tiefenbrunn, T., 32
Tokunaga, M., 114–133
Tong, W., 89–90

Topoglidis, E., 184
Torres, R., 74–84, 88, 199–200
Trabattoni, S., 229t
Trau, D., 89
Trobo-Maseda, L., 56–70
Trogler, W.C., 89
Tsai, P.-C., 22–23
Tsukahara, S., 120, 125–127, 129–130
Tung, S., 179
Turner, K.B., 21
Twigg, M.E., 50

U

Udduyasankar, U., 148
Ueda, T., 131
Ulstrup, J., 114
Unalan, I.U., 229t
Uozumi, K., 131
Ustianowski, A., 179
Uyeda, H.T., 20–21, 32

V

Vadgama, P., 181–182
Van Dongen, S.F., 89
Van Gough, D., 89–90
van Hest, J.C., 89
van Santen, R.A., 91
Varrla, E., 227–228, 229t, 231, 234–236
Verger, R., 74
Verma, M.L., 152
Vianello, F., 180–181
Volkova, A.I., 185–186
Volodkin, D.V., 93–94
von Hippel, P.H., 165
Vora, G.J., 36–37
Vrieling, E.G., 91
Vu, Q.T., 207

W

Waite, J.H., 92
Wallace, K.N., 191
Wallach, J.M., 40
Walper, S.A., 20–23, 26f, 31, 47, 49
Wan, C., 229t
Wang, C., 181–182
Wang, D., 183–184
Wang, H., 183–184
Wang, J., 114–115, 229t

Wang, J.X., 201
Wang, L., 2–5, 16, 233
Wang, L.-S., 2–16, 56–57
Wang, P., 22, 180, 185–186
Wang, Q., 114–115
Wang, X., 47, 88–89, 93–94, 102–103, 105, 107, 183–184
Wang, Y., 89–90, 232
Ward, M.D., 157
Wasalathanthri, D.P., 137–138
Wasserman, B., 183–184
Watanabe, M., 116
Wegner, D., 40–41
Wei, A., 201
Wheeler, A.R., 148
Whitesides, G.M., 182–183
Wiechelman, K., 144–146
Wieder, N.L., 183–184
Wiemann, L., 2–3
Wilker, J.J., 92
Wilkie, C.A., 159
Williams, C., 61, 199–200, 226
Willner, B., 180
Willner, I., 180
Wilson, G.S., 179
Wilson, L., 75–76
Wilson, W.C., 180
Winartasaputra, H., 200
Winkler, F.K., 74
Winter, R., 180
Witholt, B., 152
Wolosiuk, A., 89–90
Won, Y.-H., 181
Wu, C., 152
Wu, H., 89, 97
Wu, L., 89
Wu, M., 23, 25–29, 36–37, 41–44, 46–47
Wu, X., 206
Wu, Y.T., 201
Wubbolts, M.G., 152, 180
Wuhrer, M., 58

X

Xu, J., 183–184
Xu, X., 232

Y

Yacynych, A.M., 226–227
Yagonia, C.F.J., 2–3
Yamada, A., 114–133
Yáñez-Sedeño, P., 180
Yang, C., 103, 105
Yang, D., 88–110
Yang, H.-H., 226–227
Yang, J., 89
Yang, K., 227, 233
Yang, W., 114–115
Yang, W.J., 91
Yang, X., 5
Yang, Y.H., 184
Yao, Q.W., 180
Ye, C., 89
Yeh, Y.S., 89
Yi, M., 227–228, 229t
Yin, J., 229t
Yoo, Y.J., 2–3
Yoshikawa, H., 180
Yoshikawa, S., 118–119
Yu, P., 185–186
Yu, R.Q., 184
Yu, X., 138
Yun-HwaPeggy, H., 229–230

Z

Zain, R.B., 138, 142, 146–147
Zaks, A., 20
Zelikin, A.N., 89–90
Zhang, F., 185–186
Zhang, L., 93–94, 97, 103, 105–107
Zhang, M., 183–184
Zhang, Q., 232
Zhang, S., 88–110
Zhang, W., 93–94, 102–103, 107
Zhang, X., 102, 232
Zhang, Y., 114–115
Zhao, K., 233
Zhao, W.A., 182–183
Zhao, Z., 47
Zhou, Q.Z.K., 226–227
Zhou, S., 89
Zhou, Y., 22, 91
Zhu, C.-L., 226–227
Zhu, G., 180
Zhu, X., 47
Zhu, Y., 106–107
Zore, O.V., 226

SUBJECT INDEX

Note: Page numbers followed by "*f*" indicate figures, "*t*" indicate tables, and "*s*" indicate schemes.

A

Amperometric TG biosensor
  analytical recovery, added triglyceride in sera, 220–221, 221*t*
  analytic use, 220, 220*t*
  batch assay coefficient of variation (CV), 220–221, 221*t*
  evaluation, 220–221
  fabrication and response measurement, 219
  physiological concentrations, 221, 221*t*
  principle, 219
Antibody concentration, MBs, 145, 146*f*
Assay, quantification. *See* Quantification assay, enzyme–QD conjugates
2,2′-Azino-bis(3-Ethylbenzothiazoline-6-Sulfonic Acid) (ABTS) enzymatic assay, 143–144, 143*f*
α-Zr(IV) phosphate (α-ZrP) nanosheets
  entropy hypothesis, 155–156, 155*s*
  enzyme
    denaturation, 155*s*, 156
    intercalation, 157, 158*f*
    and protein, 160, 161*t*
    stabilization, 154–155, 154*s*
  exfoliation of
    characterization, 163
    powder XRD studies, 163–164
    sensitive biological molecules, 162–163
    with TBA$^+$OH$^-$, 157, 158*f*, 162–163
    zeta potential studies, 164
  layered nanomaterials, 153, 154*s*
  negatively charged proteins
    BSA–TETA, 171–172, 172*s*
    enzyme/bZrP complexes, 172–174, 173*f*, 174*s*
    GOx/Zr^{4+}/α-ZrP, activity assay for, 171, 172*s*
    metal ion mediated protein binding, 169–170, 170*f*, 171*s*
  positively charged enzymes
    CD spectra, 165–167
    complex array of interactions, 164–165
    CT activity assay, 168, 169*s*
    lysozyme activity assay, 168
    Mb and Hb peroxidase activity, 167, 167*s*
  structure of, 158–159, 158*f*
  synthesis of, 160–162
α-ZrP nanosheets. *See* α-Zr(IV) phosphate (α-ZrP) nanosheets

B

bG. *See* BioGraphene (bG)
Bicinchoninic acid (BCA) assay, 144–145, 145–146*f*
Biocatalytic microparticles
  aqueous and oil phase preparation, 12–13
  buffer preparation, 7, 7*t*
  complete protocol, 7, 8*f*
  covalent attachment, 2–3
  cross-linking, 2–3
  entrapment/encapsulation of enzymes, 2–3
  enzyme purification
    duration, 11
    flowchart, 10–11, 11*f*
    lipase preparation, 11–12
  equipment, 6
  interfacial assembly, 3–5, 5*f*, 14–15, 14*f*
  ligands capping, 5
  limitation, 2–3
  lipase, 3–5
  magnetic properties, 5
  microparticle washing, 15–16
  nanoparticle synthesis
    dissolving NaOH powder, 10
    drying process, 10
    duration, 8–10
    flowchart, 7, 9*f*

Biocatalytic microparticles (*Continued*)
  nitrogen filled vials/ethanol
    suspension, 10
  scaled up/down, 10
  washing, 10
  oil core, polymerization of, 3–5
  physical adsorption, 2–3
  preparation method, 3–5, 4f, 7
Bioconjugation, magnetic beads (MBs), 138–142
BioGraphene (bG)
  applications, 226–227
  characteristics, tunability of, 231
  disadvantages, 226
  enzyme immobilization, 226–227
  exfoliation, mechanism of, 230
  graphene exfoliation efficiency, 228, 229t
  hypothesis, 226–227
  intercalation, enzymes, 226–227
  methods
    with BSA, 233–235, 234f
    electron microscopy studies, 236–237, 237f
    HRP activity, 240–241, 241f
    protein binding, 239–240, 239f
    Raman data analysis, 235–236, 235f
    Raman spectroscopy measurements, 235
    storage and stability studies, 237
    UV-Vis measurements, 238–239, 238f
    zeta-potential measurements, 237–238, 238f
  protein-blender method, 229–230, 229s
  protein/enzyme binding, 232–233
  shear force/turbulence, 227–228
  solvent-exfoliated/surfactant-exfoliated preparations, 228
  ultrasonication process, 227–228
Biomimetic/bioinspired adhesion
  catechol chemistry, 103–105
  followed by mineralization
    surface-coating process, 108
    synthesis procedure, 108
  polyphenol chemistry, 105–106
Biomimetic/bioinspired mineralization
  biocompatible matrix formation, 91
  catechol chemistry and polyphenol chemistry, 103–106

3,4-dihydroxy-L-phenylalanine (DOPA), 91–92
followed by adhesion approach
  multienzyme system construction, 107
  synthesis procedure, 106–107
iron–DOPA complexes, 92
with LbL assembly, 97–102
metal ions, 92
physicochemical properties, 91
proteins/polyelectrolytes, 90–91
silicification process, 91
structure–function relationships, biological systems, 92
surface seggregation, 102–103
Biomimetics and bioinspiration methods
  enzyme@CaCO3 microspheres, 93–95
  rational design, 93, 93f
  structure–function relationships, biological systems, 92
  surface coating, template, 95
  technological implementation, 92–93
Biomineralization process. *See* Biomimetic/bioinspired mineralization
Bovine serum albumin (BSA), 171–172, 172s, 227–229
  characterization of, 234–235
  equipment, 233
  materials, 233
  preparation method, 233, 234f
BSA. *See* Bovine serum albumin (BSA)

C

Carboxy methyl cellulose (CMC), 123–124, 124s
Catechol chemistry
  co-precipitation method, 104
  multienzyme containing microcapsule systems, 104–105
  self-oxidative polymerization, dopamine, 104
  synthesis procedure, 104
CD spectra. *See* Circular dichroism (CD) spectra
Cerium oxide (CeO_2) NPs, enzyme-based portable biosensors
  applications, 184
  catalytic and oxygen storage/release capacity, 184

Subject Index

high oxygen mobility, 183–184
in vitro characterization, 184
optical and redox properties, 183–184
unique optical properties, 183–184
uses, 184
Chemical vapor deposition (CVD), 232
Chitosan (CHIT), NP surface chemistry
biocompatibility, 201
chemical modification, amino groups, 201
chemical structure, 201, 201f
easy availability, 201
film-forming ability, 201
gold polypyrrole (AuPPy) nanocomposites, 201
high permeability, 201
low cost, 201
mechanical strength, 201
non-toxicity, 201
poly (indole-5-carboxylic acid) (Pin5COOH), 201
Chymotrypsin (CT) activity assay, 168, 169s
Circular dichroism (CD) spectra
baseline correction, 166–167
characteristics, 165–166
native-like structure, 165–166
recording control spectra, 165–166
Colorimetric enzyme-based assays
advantages, 186
biosensor concept and detection mechanism, 185–186, 185f
CeO_2-enzyme paper, 187
CeO_2 NP sensors, 186–187
CeO_2-silane paper, 187
reagents, 187
sensing elements, 186
Confocal laser scanning microscopy (CLSM), 96

E

Electron microscopy studies, 236–237, 237f
Enzyme activity sensors
equipment, 34
kinetics measurements, 36
materials, 34–36
peptide labeling, 34–35
peptide precleaving, 35
QD–peptide construct, 35–36
spectral characterization, 36

Enzyme-based portable biosensors
CeO_2 NPs
applications, 184
catalytic and oxygen storage/release capacity, 184
high oxygen mobility, 183–184
in vitro characterization, 184
optical and redox properties, 183–184
unique optical properties, 183–184
uses, 184
color analysis, 191
colorimetric enzyme-based assays
advantages, 186
biosensor concept and detection mechanism, 185–186, 185f
CeO_2-enzyme paper, 187
CeO_2 NP sensors, 186–187
CeO_2-silane paper, 187
reagents, 187
sensing elements, 186
detection and measurement procedure, 188–190, 189f
interferences testing, 191
NPs
biosensing design, 179–180
color changing properties and reactivity, 190
conjugation, 178–179
enzyme immobilization, 180–181, 182f, 191
low-cost and portable devices, 181–183
surface characteristics, 190
Enzyme@capsule nano/microsystems
bioinspiration (see Biomimetic/bioinspired mineralization)
biomimetic/bioinspired chemistries, 90–93
biomineralization (see Biomimetic/bioinspired mineralization)
capsules, 89
catalytic activity and stability, enzyme, 97, 98t
clean energy products, 88
covalent cross-linking, 88
design and construction, 89–90
encapsulation, 88
entrapment, 88

Enzyme@capsule nano/microsystems
  (*Continued*)
 enzyme location, 96
 formation process
  charge (pI), 94
  hydrophobicity, 94
  side chains, 94
  size, 94
 hard-templating strategies, 89
 higher stirring rate, 94
 intercalation, 88
 LbL assembly, oppositely charged
  polyelectrolytes, 89–90
 pharmaceutical/fine/commodity
  chemicals, 88
 protein co-precipitation, 93–95
 SEM/TEM, 96
 soft-templating methods, 89
 surface grafting, 88
 template-free methods, 89
 template removal, EDTA/dilute HCl
  treatment, 95
 verification, formation process,
  95–96
 water-insoluble carbonates, 90
Enzyme encapsulation.
  See Enzyme@capsule nano/
  microsystems
Enzyme immobilization
 CeO_2 NPs
  applications, 184
  catalytic and oxygen storage/release
   capacity, 184
  high oxygen mobility, 183–184
  in vitro characterization, 184
  optical and redox properties, 183–184
  unique optical properties, 183–184
  uses, 184
 CeO_2-silane paper, 187
 NPs, 180–181, 182*f*, 191
Enzyme-intercalated oxide layers
 cellulase concentration, 120–122, 123*f*
 cellulase-FT composites, 120–122, 121*f*
 cellulase intercalated FT layers, 120–122,
  122*f*
 with micrometric dimensions
  colloidal solutions, 116
  KFT powder, 116–117, 117*f*
  neutralization and dilution, 117–118,
   118*f*
  protonated powders (HFT), 116–118,
   117*f*
  tetrabutylammonium hydroxide,
   117–118, 118*f*
  XRD patterns, 116–117, 117*f*
 nanohybrids, enzymes and anionic oxide
  layers, 114–115, 115*f*
 with nanometric dimensions, 118–119,
  119*f*
 origin, 115
 zeta potential, pH dependence of,
  120–122, 120*f*
Enzyme kinetics, MBs
 biomolecules, 139
 EDC–NHSS chemistry, 139
 optimal bioanalytical performance,
  138–139
 streptavidin-coated, 139
Enzyme–nanoparticle constructs
 activity and kinetic parameters, 20
 environmental optimization, 20
 inorganic colloidal NPs, 20–21
Enzyme surfaces
 chemical amination, 60
 dextran–aldehyde polymers, 58,
  59–60*f*
 enzyme active center, 62
 hierarchical self-assembly, 56–57
 hyperhydrophilic layer, 57
 LbL method, 56–57
 limitations, 56
 Lys, 60
 massive PEGylation, 57
 physical amination, 61, 61*f*
 stabilizing effects, 56–57, 61–62
 thermal inactivation
  CALB derivatives, 69*f*
  endoxylanase derivatives, 70*f*
  RML derivatives, 68*f*
  Thermomyces lanuginosa lipase
   derivatives, 69*f*
Exfoliation
 mechanism of, 230
 α-ZrP nanosheets
  characterization, 163
  powder XRD studies, 163–164

sensitive biological molecules, 162–163
with TBA$^+$OH$^-$, 157, 158f, 162–163
zeta potential studies, 164

F

Few layer graphene (FLG), 233
Fluorescence
  of enzyme sensor, 44f
  QD, 41, 43
  Tecan plate reader, 43
  UV light irradiation, 126–127
Förster resonance energy transfer (FRET)
  mechanism, 23, 41, 41f

G

Glycerol kinase (GK)
  applications (see amperometric TG biosensor)
  combined assay, 202–215
  covalent co-immobilization
    Pin5COOH/AuPPynano-composite onto Au electrode (see Gold nanoparticles-polypyrrole-polyindole carboxylic acid (AuPPy-Pin5COOH))
    ZnONPs/CHIT electrodeposited (see Zinc oxide nanoparticles (ZnONPs), Pt electrode; ZnONPs-CHIT composite, Pt electrode)
  enzyme preparation mixture, 202
  kinetic properties
    optimum pH, 215, 216f, 218t
    response time, 217, 217f, 218t
    storage stability, 218, 218t
    temperature, 216–217, 216f, 218t
    working range/linearity, 217–218, 218t
Glycerol-3-phosphate oxidase (GPO)
  applications (see amperometric TG biosensor)
  combined assay, 202–215
  covalent co-immobilization
    Au electrode (see Gold nanoparticles-polypyrrole-polyindole carboxylic acid (AuPPy-Pin5COOH))
    Pt electrode (see Zinc oxide nanoparticles (ZnONPs), Pt electrode; ZnONPs-CHIT composite, Pt electrode)
  enzyme preparation mixture, 202
  kinetic properties
    optimum pH, 215, 216f, 218t
    response time, 217, 217f, 218t
    storage stability, 218, 218t
    temperature, 216–217, 216f, 218t
    working range/linearity, 217–218, 218t
Gold nanoparticles-polypyrrole-polyindole carboxylic acid (AuPPy-Pin5COOH)
  bare Pt electrode, SEM, 208–211, 209–210f
  chemical reaction, enzyme immobilization, 208, 209f
  CHIT/electrode, SEM, 208–211, 209–210f
  co-immobilization, lipase, GK, and GPO, 208
  confirmation, enzyme immobilization, 208–215
  construction, 206
  cyclic voltammograms (CV), 214–215, 214–215f
  EIS spectra, 211–212, 211–212f
  electrodeposition, 207–208
  enzymes/ZnONPs/CHIT/electrode, SEM, 208–211, 209–210f
  FTIR spectra, 212–214, 213f
  nanocomposites preparation, 207
  TEM characterization, 206
  ZnONPs/CHIT/electrode, SEM, 208–211, 209–210f

H

HRP activity, 240–241, 241f

I

Immunoassay
  dual-labeled MBs, 146–147
  in microfluidic systems, 148
  protein analyte, 146–147
  protein capture, 148
  protein detection systems, 148
  selectivity and sensitivity, 147
  ultra-low detection limits, 147–148

K

Kinetics, QD-enzyme constructs
  linear regression analysis, 31
  4-nitrophenol concentration, 31, 31f
  plate reader, absorbance measurement, 30
  product calibration curve, 31
  substrate dilution table, 29, 29t, 30f

L

Layer-by-layer (LbL) assembly, biomimetics/bioinspiration
  EDTA treatment, 102
  protamine–titania capsules, 97
  surface-coating process, 102
  synthesis procedure, 97–102
Lipase
  applications (see amperometric TG biosensor)
  combined assay, 202–215
  covalent co-immobilization
    Au electrode (see Gold nanoparticles-polypyrrole-polyindole carboxylic acid (AuPPy-Pin5COOH))
    Pt electrode (see Zinc oxide nanoparticles (ZnONPs), Pt electrode; ZnONPs-CHIT composite, Pt electrode)
  enzyme preparation mixture, 202
  kinetic properties
    optimum pH, 215, 216f, 218t
    response time, 217, 217f, 218t
    storage stability, 218, 218t
    temperature, 216–217, 216f, 218t
    working range/linearity, 217–218, 218t
Lysozyme (Lys) activity assay, 168

M

Magnetic beads (MBs)
  ABTS assay, 143–144
  BCA assay, 144–145, 145–146f
  binding kinetics
    biomolecules, 139
    EDC–NHSS chemistry, 139
    optimal bioanalytical performance, 138–139
    streptavidin-coated, 139
  detection antibodies (Ab2), 138
  dual-labeled
    streptavidin MBs, 141–142, 141f
    tosyl-activated MBs, 139–141, 140f
  electrochemical detection, 137–138
  electrochemiluminescence, 137–138
  immunoassays, 136–137, 146–148
  organic functional groups, 136–137, 137f
  preconcentration factors, 137–138
  Smith assay, 144–145
  surface chemistries, 137–138
  surface plasmon resonance, 137–138
Michaelis–Menten kinetics, 3–5, 32, 33f, 48
Myoglobin and hemoglobin peroxidase activity, 167, 167s

N

Nanoparticles (NPs)
  enzyme-based portable biosensors
    biosensing design, 179–180
    color changing properties and reactivity, 190
    conjugation, 178–179
    immobilization of enzymes, 180–181, 182f, 191
    low-cost and portable devices, 181–183
    surface characteristics, 190
  synthesis
    dissolving NaOH powder, 10
    drying process, 10
    duration, 8–10
    flowchart, 7, 9f
    nitrogen filled vials/ethanol suspension, 10
    scaled up/down, 10
    washing, 10
NPs. See Nanoparticles (NPs)

P

PEGylation
  chemically aminated enzymes
    Rhizomucor miehei lipase, 62–64
    Thermomyces lanuginosa lipase, 64–65
  enzymes coated with polymers
    Bioxilanase L Plus, 66–68
    Candida antarctica B lipase, 65–66
Peptide labeling
  dual-labeled peptides, 37–38, 38f
  optimum peptide design, 36–37

peptide purification, quantification and storage, 38–39
polyproline spacer, 36–37
single cysteine labeling, 37
Peptide precleaving, 39–40
Photoreaction model, 126–127, 126f
Poly (indole-5-carboxylic acid) (Pin5COOH) and AuPPy nanoparticles
bare Pt electrode, SEM, 208–211, 209–210f
chemical reaction, enzyme immobilization, 208, 209f
CHIT/electrode, SEM, 208–211, 209–210f
co-immobilization, lipase, GK, and GPO, 208
confirmation, enzyme immobilization, 208–215
cyclic voltammograms (CV), 214–215, 214–215f
EIS spectra, 211–212, 211–212f
electrodeposition, 207–208
enzymes/ZnONPs/CHIT/electrode, SEM, 208–211, 209–210f
FTIR spectra, 212–214, 213f
nanocomposites preparation, 207
ZnONPs/CHIT/electrode, SEM, 208–211, 209–210f
Polyphenol chemistry
surface-coating process, 106
synthesis procedure, 105
Potein-blender method, 229–230, 229s

Q

QD–peptide construct
agarose gel, 41, 42f
characterization, 41–42
data analysis, 47–48, 48f
DLS, 42
enzyme activity sensors, 23–24, 34–48
fixed enzyme experiments, 45–46
fixed substrate experiments, 46–47
formation, 40–41
hydrodynamic radius, 42
quantification assay, 22–23
spectral characterization, 43–45, 44f
terminal FRET acceptors, 41, 41f

Quantification assay, enzyme–QD conjugates
bioremediation/biosafety applications, 22–23
data analysis
Michaelis–Menten curves, 32, 33f
paraoxon hydrolysis, 32, 32f
SLOPE and RSQ functions, 32, 33f
enzyme assembly, 25–28
enzyme catalytic enhancements, 22
enzyme recycling, 22
kinetic data, 29–31
materials
equipment, 24
reagents, 24–25
sequential variable modifications, 22
Quantum dots (QD)
binding, metal ion complexation, 21
enzyme assembly
calibration curve, 31, 31f
gel electrophoresis, 25, 26f
hydrophilic ligand structure, 25–29, 26f
linear regression analysis, 31
phosphotriesterase (PTE), 25–29
enzyme's reaction kinetics, 21
food technology, 23–24
FRET, 23
medical applications, 23–24
optical properties, 21
proteases, 23–24
reversible interactions, 21

R

Raman data analysis, 235–236, 235f
Raman spectroscopy measurements, 235
Relative fluorescence unit (RFU), 126–127, 127f

S

Serum triglycerides
atherosclerosis, 198–199
biocatalyst preparation methodology, 199–200
color reaction, 199
communal enzyme-particle approach, 200
covalent binding, 200

Serum triglycerides (*Continued*)
  enzymatic colorimetric method, 199–200
  metabolism, 198–199
  TG levels, 198–199
Smith assay, 144–145, 145–146f
Spectrophotometry, 24, 36
Storage and stability studies, 237
Streptavidin-coated magnetic beads, 141–142, 141f
Surface seggregation, biomimetics/bioinspiration
  PAH/GA-titania-enzyme@CaCO$_3$, 103
  P-enzyme@CaCO3 microspheres, 103
  surface-coating process, 103
  synthesis procedure, 102–103
Synergistic functions
  fabrication of (*see* enzyme-intercalated oxide layers)
  layered metal oxides, 115
  magnetic application
    enzyme and FT layers, 130–131, 130f
    HRP/titanate/MB nanohybrids, 131–132
    nanometric undoped titanate, 131–132
  three-dimensional framework, 114
  titanate layers
    anti-UV light stability, 125
    biorecognition process, Eu-doped, 129, 129f
    cellulase and FT layers, 123–124, 124f
    CMC, decomposition of, 123–124, 124s
    nanometric, 125
    UV light irradiation
      photoreaction model, 126–127, 126f
      RFU, 126–127, 127f
      visible-light-driven enzymatic activity, 127–129, 128f

T

Tosyl-activated MBs, 139–141
Triethylenetetramine (TETA), 171–172, 172s

U

Ultrasonication process, 227–228
UV–Vis measurements, 238–239, 238f

Z

Zeta-potential measurements, 237–238, 238f
Zinc oxide nanoparticles (ZnONPs), Pt electrode
  characterization, 203
  preparation, 203
  shape and size, 203, 204f
ZnONPs-CHIT composite, Pt electrode
  deposition, 205
  enzyme immobilization chemistry, 205, 206f
  preparation, 204–205

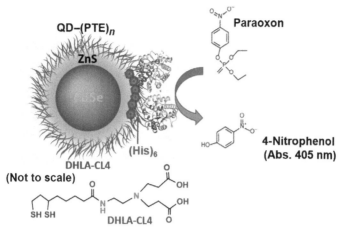

S.A. Díaz et al., Figure 1 Schematic of enzyme–QD assembly. In this example, the enzyme is PTE and 525 nm CdSe/ZnS QDs have been rendered colloidally stable using the zwitterionic ligand DHLA-CL4. PTE is attached to the ZnS-rich surface of the QD through metal coordination with the His_6 engineered to the N-terminus end of the enzyme. PTE catalyzes the hydrolysis of paraoxon to 4-nitrophenol whose formation can be monitored by measuring the absorbance at 405 nm over time. Adapted from Breger, Ancona, et al. (2015).

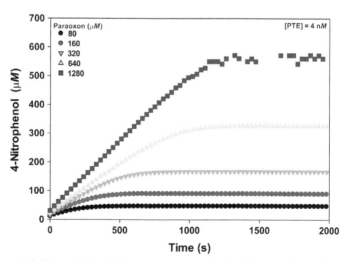

S.A. Díaz et al., Figure 6 Typical progress curves obtained from the hydrolysis of paraoxon catalyzed by PTE. The absorbance values have been converted to product concentration values using a standard curve. Absorbance values that exceeded the detection limits of the plate reader were discarded.

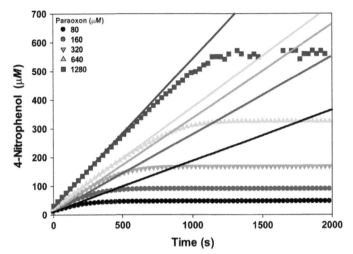

S.A. Díaz et al., Figure 7 Representation of the linear fits utilized to determine the initial velocities from the data presented in Fig. 6. Only the initial linear portions of the progress curves for each substrate concentration should be utilized. These correspond to time windows of circa 0 to 150, 200, 250, 500, and 500 s for the 80, 160, 320, 640, and 1280 μM paraoxon substrate samples utilized, respectively, in this figure.

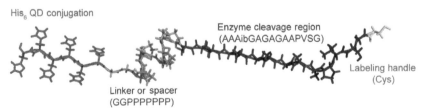

S.A. Díaz et al., Figure 9 Schematic of elastase substrate peptide which binds to the QD. The four partitions mentioned in Section 3.2 are noted in the figure.

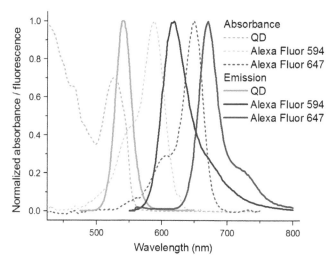

S.A. Díaz et al., Figure 10 Spectra of fluorescent components. Absorbance (dashed lines) and Emission (full lines) are normalized to a maximum peak to demonstrate spectral overlap.

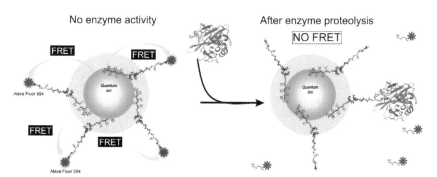

S.A. Díaz et al., Figure 11 Schematic of QD–peptide enzyme activity sensor. Left: Before the presence of enzyme activity the substrate presents terminal FRET acceptors. Right: In the presence of enzyme activity the substrate is cleaved, the acceptor then diffuses away from the QD. The extended distance reduces the FRET efficiency, increasing QD emission, and decreasing the acceptor emission.

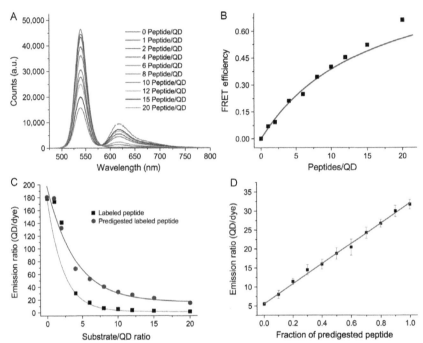

S.A. Díaz et al., Figure 13 Calibration curves of enzyme sensor. (A) Fluorescence spectra of sensor as the number of peptide/QD increases. (B) FRET efficiency of the sensor as a function of the peptide/QD ratio. Squares are experimental data while the line corresponds to a FRET fit. It is important to note that at higher ratios deviation from the FRET formalisms begins to increase. (C) Emission ratios as a function of the substrate/QD ratio. It is important to note that a fully cleaved system is not equivalent to a system that has no dye in solution. Similarly, the curves demonstrate that at low substrate/QD ratios, the system is too volatile for confident measurements. The lines are simply to guide the eye, though proper numerical fits can be achieved. (D) Emission ratio of a 10 substrate/QD construct as a function of the fraction of predigested peptide.

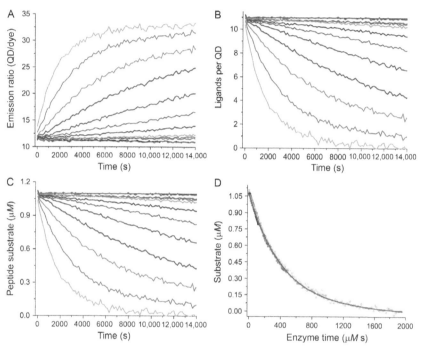

S.A. Díaz et al., Figure 14 Step-by-step transformations of a fixed substrate enzyme sensor experiments. (A) Raw emission ratio curves. (B) Using the calibration curves and baseline correction to transform emission ratios into remaining ligands per QD. (C) Multiplication by QD concentration to obtain peptide substrate consumption progress curves. (D) Enzyme time peptide substrate consumption progress curves. The points are the overlying data points from the curves in part C. The blue line is the corresponding fit from Eq. (3).

K. Kamada et al., Figure 3 Photographs of colloidal solutions of protonated Fe-doped titanates during (A) exfoliation with a tetrabutylammonium hydroxide and (B) after the exfoliation followed by neutralization and dilution (pH 8.3).

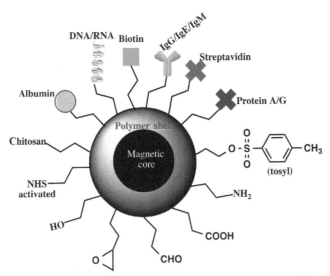

B.A. Otieno et al., Figure 1 Variety of commercially available functionalized MBs with coatings of either organic functional groups to attach biomolecules or particular biomolecules that can bind specific moieties.

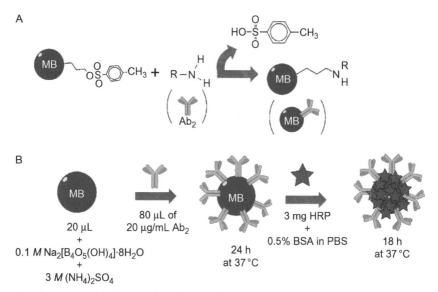

B.A. Otieno et al., Figure 2 (A) The covalent attachment of antibodies to the tosyl functionalized magnetic particles. The tosyl groups act as leaving groups for surface amine groups present on antibodies for covalent attachment. (B) The complete conjugation protocol for both the attachment of antibodies as well as HRP enzyme labels.

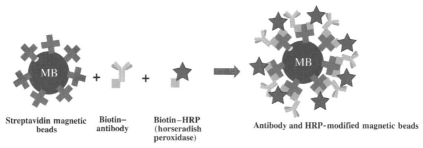

B.A. Otieno et al., Figure 3 The noncovalent attachment of biotin–antibodies and biotin–horseradish peroxidase labels to the surface of streptavidin-coated MBs.

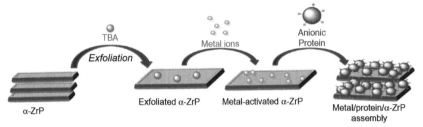

I.K. Deshapriya and C.V. Kumar, Figure 3 Cation-mediated binding of anionic enzymes to anionic α-ZrP nanosheets.

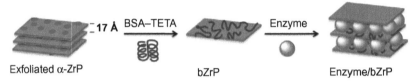

I.K. Deshapriya and C.V. Kumar, Figure 4 Binding of negatively charged enzymes/protein into the galleries of exfoliated α-ZrP with the aid of cationized BSA.

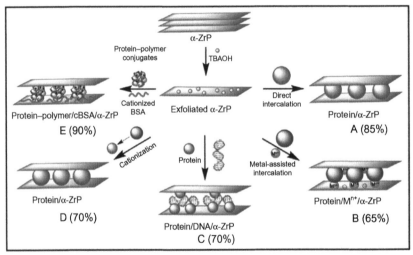

I.K. Deshapriya and C.V. Kumar, Scheme 1 Approaches (A)–(E) for controlling enzyme intercalation into α-ZrP galleries. Exfoliation of the nanosheets with tetra (n-butylammonium) hydroxide (TBA$^+$OH$^-$) followed by: (A) direct binding of positively charged enzymes to the anionic nanosheets, (B) metal ion-mediated binding of anionic enzymes to the anionic nanosheets, (C) Hb-mediated binding of anionic DNA/RNA, (D) cationization of anionic enzymes to promote their binding to the anionic nanosheets, and (E) design of protein-glues for the binding of anionic enzyme–polymer conjugates to the anionic nanosheets. A comparison of the extent of retention of hemoglobin peroxidase-like activity is given in the brackets for each of these approaches (A)–(F).

Hypothesis

Stabilize the native state and destabilize the denatured state (SNSDDS)!

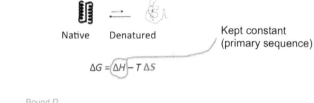

Intercalation of the enzyme between the nanosheets might lower the entropy of the denatured state and hence stabilize the intercalated enzyme!

I.K. Deshapriya and C.V. Kumar, Scheme 2 Schematic diagram showing the "Entropy" hypothesis. A general approach to stabilize enzymes by lowering the conformational entropy of the denatured state. The free energy gap between the native and denatured states may be increased, which results in improved stability of the enzyme.

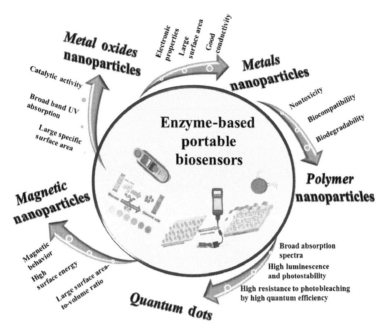

A. Karimi et al., Figure 1 Overview of the various NPs systems and their properties for enzyme-based portable biosensors.

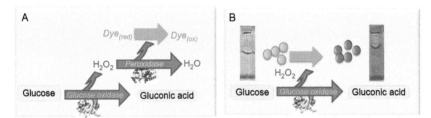

A. Karimi et al., Figure 2 Schematic representation of the working principle of colorimetric enzyme-based assays for the detection of glucose: (A) conventional assay involving the use of HRP and an organic dye. (B) CeO_2 NPs-based assay for the detection of glucose. This principle has been demonstrated on paper surfaces with both the enzyme and the NPs stabilized on paper to create a reagentless enzyme assay (Ornatska et al., 2011).

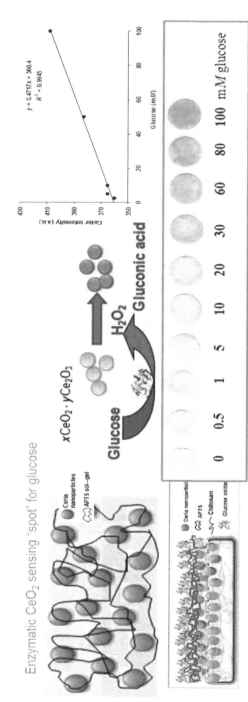

A. Karimi et al., Figure 3 CeO$_2$–GOX paper-based biosensor and colorimetric response to glucose concentrations ranging from 0 to 100 m*M* glucose.

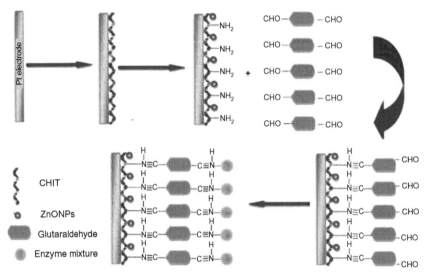

V. Aggarwal and C.S. Pundir, Figure 3 Schematic representation of ZnONPs/CHIT electrode preparation and immobilization of enzymes on it. Step 1 + Step 2: Uniform spreading of chitosan solution containing ZnONPs onto Pt electrode; Step 3: Activation of CHIT/ZnONPs/PtE; Step 4: Covalent co-immobilization of mixture of lipase, GK, and GPO onto activated electrode. *Reprinted from Narang and Pundir (2011), with permission from Elsevier.*

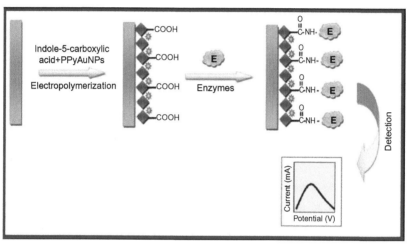

V. Aggarwal and C.S. Pundir, Figure 5 Schematic representation of AuPPy/Pin5COOH electrode preparation and chemical reactions involved in covalent co-immobilization of enzymes (◦, AuPPy; ◆, Pin5COOH). *Reprinted from Narang et al. (2012), with permission from Elsevier.*

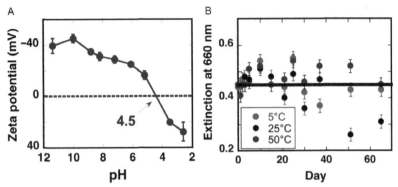

C.V. Kumar and A. Pattammattel, Figure 4 (A) Change in zeta potential of BioGraphene with pH. (B) Absorbance measurements show stability over 2 months at 5°C or 50°C.

C.V. Kumar and A. Pattammattel, Figure 5 HRP binding to the nanosheets of BioGraphene for sensing or biocatalysis applications.

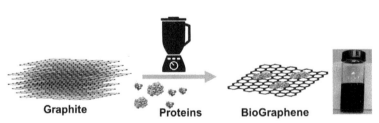

C.V. Kumar and A. Pattammattel, Scheme 1 Production of BioGraphene using proteins in a kitchen blender.